A Text Book Of

AUTOMOBILE COMPONENT DESIGN

(22558)

Semester - V

THIRD YEAR DIPLOMA IN AUTOMOBILE ENGINEERING GROUPS

As Per MSBTE's 'I' Scheme Syllabus

Dilip M. Kupade
DAE. BE (Mechanical), ME (Design Engg.)
Lecturer in Mechanical Engg. Dept.,
ZES' Zeal Polytechnic, Narhe,
Pune.

Shivaji S. Aher
B.E. Mechanical Engineering
M.E. (Machine Design)
H.O.D. (Automobile Department)
Shivajirao S. Jondhle Polytechnic, Asangaon.

N4523

Automobile Component Design (Automobile Engg. Gr.) (Sem. V)　　　　ISBN 978-93-89108-43-9

First Edition　:　June 2019
©　　　　　　:　**Authors**

Published By :
NIRALI PRAKASHAN
Abhyudaya Pragati, 1312, Shivaji Nagar
Off J.M. Road, PUNE – 411005
Tel - (020) 25512336/37/39, Fax - (020) 25511379
Email : niralipune@pragationline.com

➢ DISTRIBUTION CENTRES

PUNE

Nirali Prakashan　:　119, Budhwar Peth, Jogeshwari Mandir Lane, Pune 411002, Maharashtra
(For orders within Pune)　　Tel : (020) 2445 2044, Fax : (020) 2445 1538; Mobile : 9657703145
　　Email : bookorder@pragationline.com, niralilocal@pragationline.com

Nirali Prakashan　:　S. No. 28/27, Dhyari, Near Pari Company, Pune 411041
(For orders outside Pune)　　Tel : (020) 24690204 Fax : (020) 24690316; Mobile : 9657703143
　　Email : dhyari@pragationline.com, bookorder@pragationline.com

MUMBAI

Nirali Prakashan　:　385, S.V.P. Road, Rasdhara Co-op. Hsg. Society Ltd.,
　　Girgaum, Mumbai 400004, Maharashtra; Mobile : 9320129587
　　Tel : (022) 2385 6339 / 2386 9976, Fax : (022) 2386 9976
　　Email : niralimumbai@pragationline.com

➢ DISTRIBUTION BRANCHES

JALGAON

Nirali Prakashan　:　34, V. V. Golani Market, Navi Peth, Jalgaon 425001,
　　Maharashtra, Tel : (0257) 222 0395, Mob : 94234 91860
　　Email : niralijalgoan@pragationline.com

KOLHAPUR

Nirali Prakashan　:　New Mahadvar Road, Kedar Plaza, 1st Floor Opp. IDBI Bank
　　Kolhapur 416 012, Maharashtra. Mob : 9850046155
　　Email : niralikolhapur@pragationline.com

NAGPUR

Nirali Prakashan　:　Above Maratha Mandir, Shop No. 3, First Floor,
　　Rani Jhanshi Square, Sitabuldi, Nagpur 440012, Maharashtra
　　Tel : (0712) 254 7129

DELHI

Nirali Prakashan　:　4593/15, Basement, Agarwal Lane, Ansari Road, Daryaganj
　　Near Times of India Building, New Delhi 110002 Mob : 08505972553
　　Email : niralidelhi@pragationline.com

BENGALURU

Nirali Prakashan　:　Maitri Ground Floor, Jaya Apartments, No. 99, 6th Cross, 6th Main,
　　Malleswaram, Bangaluru 560 003, Karnataka
　　Mob : +91 9449043034
　　Email: niralibangalore@pragationline.com

niralipune@pragationline.com | www.pragationline.com

Also find us on www.facebook.com/niralibooks

Dedicated to

Automobile Engineers

Preface ...

It gives us great pleasure to present first edition of the text book on "**Automobile Component Design**" for Third Year Diploma I-Scheme students. This is applied technology subject, which requires knowledge of Mechanism, Strength of Materials, Material science, Manufacturing Process and Mechanical Engineering Drawing.

This Text book covers fundamental Principles of Machine Design applied to automobile components and also revised syllabus prescribed by "Maharashtra State Board of Technical Education".

In this book, both Theoretical concepts and Numerical illustrations have been appropriately clear the subject matter. Typical questions have been added at the end of each chapter to enhance the utility of this book.

We express our deep gratitude towards Mr. Jignesh Furia, Publisher of Nirali Prakashan.

We are also thankful to Mrs. Manasi Pingle, Mr. Ilyas Shaikh, Mrs. Deepa Sawant and staff members of Nirali Prakashan for bringing out this book in the shortest possible time.

Special thanks to our family members for their co-operation.

The suggestions and errors pointed out by the reader are welcome.

Authors

Syllabus ...

1. Fundamentals of Automobile Component Design   (Hrs. 10; Marks 12)

1.1   Component Design - Concept

1.2   Modes of Failure of Different Automotive Components

1.3   Basic Requirements of Automobile Components

1.4   Basic Automobile Component Design Procedure

1.5   Use of Standards in Component Design

1.6   Preferred Numbers

1.7   Ergonomic Considerations in Component Design

1.8   Aesthetic Considerations in Component Design

2. Stresses in Automobile Components   (Hrs. 12; Marks 12)

2.1   Normal and Shear Stresses, Crushing Stress, Bearing Pressure, Torsional and Bending Stresses, Principal Stresses, Variable Stresses, Impact Stresses, Resilience

2.2   Stress-Strain Diagram and its Uses

2.3   Working Stress, Factor of Safety, Selections of FOS

2.4   Theories of Failure Under Static Loading - Maximum Principal or Normal Stress Theory, Maximum Shear Stress Theory, Maximum Distortion Energy Theory

2.5   Fatigue, Endurance Limit, FOS for Fatigue Loading, S-N Curve

2.6   Stress Concentration, Causes and Remedies

2.7   Load Factor, Service Factor and Their Applications

3. Design of Chassis Components   (Hrs. 16; Marks 16)

3.1   Function of Tie Rod, Materials for Tie Rod with Justification and Design of Tie Rod

3.2   Function of Clutch, Material for Friction Lining with Justification, Design of Disc Clutch and Multi-plate Clutch Considering Uniform Wear Condition

3.3   Function of Propeller Shaft, Design of Propeller Shaft Including Universal Coupling

3.4   Function of Semi-elliptical Leaf Spring, Materials of Leaf Spring with Justifications and Design of Semi-elliptical Leaf Spring

4. Design of Engine Components   (Hrs. 18; Marks 16)

4.1   Function of Cylinder Block, Materials for Cylinder Block with Justifications. Design of Bore Diameter; Bore Length and Thickness of Cylinder Wall

4.2   Function of Piston, Materials with Justification and Design of Piston and Piston Pin

4.3   Function of Connecting Rod, Materials with Justification and Design of Connecting Rod

4.4   Function of Rocker Arm, Materials with Justification and Design of Rocker Arm (For Rectangular Cross Section Only)

4.5   Function of Valve Spring, Materials with Justification and Design of Valve Spring

4.6   Function of Push Rod, Material with Justification and Design of Push Rod

5. Design of Axles   (Hrs. 08; Marks 14)

5.1   Function of Front Axle, Material with Justification and Design of Front Axle

5.2   Function of Rear Axle, Material for Rear Axle with Justification and Design of Full Floating Rear Axle

✍ ✍ ✍

Contents

☝☝☝

Chapter 1

FUNDAMENTALS OF AUTOMOBILE COMPONENT DESIGN

Weightage of Marks = 12, Teaching Hours = 10

Syllabus

1.1   Component Design - Concept

1.2   Modes of Failure of Different Automotive Components

1.3   Basic Requirements of Automobile Components

1.4   Basic Automobile Component Design Procedure

1.5   Use of Standards in Component Design

1.6   Preferred Numbers

1.7   Ergonomic Considerations in Component Design

1.8   Aesthetic Considerations in Component Design

About this Chapter

After reading this chapter, students will be able to :

- Describe mode(s) of failure of the given automobile component/s with sketches.

- List the standards used in design of the given automobile component/s.

- Explain use of preferred numbers in designing the given automobile component.

- Explain effect of ergonomics on the given automobile component design.

- Explain effect of aesthetics on the given automobile component design.

1.1 COMPONENT DESIGN - CONCEPT

- Automobile component design is applied technology subject which requires knowledge of mechanisms, strength of materials, material sciences, manufacturing processes and mechanical engineering drawing.

- The aim of this subject is to create a new machine and to improve the existing ones. The better machine or new machine is one which is more economical in overall cost of production and operation.

- Machine design is defined as the "use of scientific principles, technical information and imagination in the description of mechanical system to perform specific functions with maximum economy and efficiency".

1.1.1 Classification of Design   (S-17)

The machine design is classified as follows :

1. **Development design :** This type of design is to modify the existing designs into a new idea by adopting a new material or different methods of manufacture.

2.   New design : This type of design needs research work, technical ability and creative thinking.

Methods Used in New Design :

(a) System design (W-15) : It is the design of any complex mechanical system like a motor car which will satisfy the specified requirement of system. System design could be seen as the application of systems theory to product development.

(b) Component design : It is the design of any component of the mechanical system like piston, crankshaft, connecting rod etc.

(c) Optimum design : It is the best design for the given objective function under the specified constraints. It is achieved by minimizing the undesirable effects.

(d) Computer aided design : This type of design depends upon the use of computer systems to assist in the creation, modification, analysis and optimization of a design.

(e) Product design (W-15) : This type of design creates end product.   OR

It is the process of creating a new product to be sold by a business to customers. It is essentially the efficient and effective generation and development of the ideas through a process that leads to new products.

1.2 MODES OF FAILURE OF AUTOMOBILE COMPONENTS   (S-16)

Following are the types of failure :

1.	Brinnelling	2.	Force and/or temperature induced elastic deformation
3.	Yielding	4.	Brittle fracture
5.	Ductile rupture	6.	Wear
7.	Fatigue	8.	Corrosion
9.	Buckling	10.	Creep
11.	Impact	12.	Fretting

1.   Brinnelling : This failure occurs when the static forces between two curved surfaces in contact result in local yielding of one or both mating members to produce a permanent surface discontinuity of significant size.

2.   Force and/or temperature induced elastic deformation : This failure occurs when the elastic deformation in a machine member is brought about by the imposed operational loads or temperatures, becomes great enough to interfere with the ability of the machine to perform satisfactorily its intended function.

3.   Yielding : This failure occurs when the plastic deformation in a ductile material is brought about by the imposed operational loads or motions. And it becomes great enough to interfere with the ability of the machine to perform satisfactorily its intended function.

4.   Brittle fracture : This failure occurs when the elastic deformation in a machine part that exhibits brittle behaviour is carried out to the extreme so that, the primary interatomic bonds are broken and the member separates into two or more pieces.

5.   Ductile rupture : This failure occurs when the plastic deformation in machine part that exhibits ductile behaviour is carried out to the extreme, so that the member separates into two pieces.

6.   Wear : Wear is an undesired cumulative change in dimension brought about by the general removal of discrete particles from contacting surfaces in motion. Usually sliding, predominantly as a result of mechanical action.

7.   Fatigue : Fatigue failure is general term given to the sudden and catastrophic separation of a machine part into two or more pieces, as a result of the application of fluctuating loads or deformations over a period of time. Failure takes place by the imitation and propagation of a crack until it becomes unstable, and propagates suddenly to failure. The loads and deformations that cause failure by fatigue are typically far below the static failure levels.

8. Corrosion : This failure implies that, a machine part is rendered incapable of performing its intended function because of the undesirable deterioration of the material as a result of chemical or electrochemical interaction with the environment. Corrosion often interacts with other failure modes such as wear or fatigue.

9. Buckling : This failure occurs when, because of critical combination of magnitude and/or point of load application, together with the geometrical configuration of a machine member, the deflection of the member suddenly increases greatly with only a slight increase in load. This non-linear response results in buckling failure.

10. Creep : This failure results when the plastic deformation in a machine member occurs over a period of time under the influence of stress and temperature until the accumulated dimensional changes interfere with the ability of the machine part to perform satisfactorily its intended function.

11. Impact : This failure occurs when a machine member is subjected to monostatic loads that produce the part stresses or deformations of such magnitude that, the member no longer is capable of performing its function. The failure is brought about by the interaction of stress or strain generated by the dynamic or suddenly applied loads, which may induce local stresses and strains many times greater than would be induced by static application of the same loads.

12. Fretting : Fretting action may occur at the interface between any two solid bodies whenever they are pressed together by a normal force and subjected to small amplitude cyclic relative motion with respect to each other. Fretting usually takes place in joints that are not intended to move out, because of vibrational loads or deformations, experience minute cyclic relative motions.

1.3 CONSIDERATIONS IN AUTOMOBILE DESIGN (W-16, 18; S-18)

Q.1. *Explain considerations in automobile design.* (W-17)

Following are the general considerations in designing a machine element :
(a) Type of load and stresses caused by load.
(b) Motion of the parts or kinematics of the machine.
  The motion of the parts may be rectilinear, reciprocating or curvilinear motion which includes rotary, oscillatory and simple harmonic, constant velocity, constant or variable acceleration etc.
(c) **Selection of materials :** It is essential that a designer should have a thorough knowledge of the properties of materials and their behaviour under working conditions.
(d) Shape and size of parts. (e) Frictional resistance and lubrication.
(f) Convenient and economical features. (g) Use of standard parts.
(h) Safety of operation. (i) Workshop facilities.
(j) Number of machines to be manufactured. (k) Cost of construction.
(*l*) Assembling.

1.3.1 Selection of Materials for Automobile Component

It is most difficult to select proper material for engineering purpose. The material which gives desired objectives at minimum cost is best one. The following factors should be considered while selecting the materials :
1. Availability of the materials. 2. Suitability of the materials for the working conditions in service.
3. The cost of the materials.

1.3.3.1 Mechanical Properties

The mechanical properties of the metals are those which are associated with the ability of the material to resist mechanical forces and load.
1. **Strength** (S-16) **:** It is the ability of a material to resist the externally applied forces without breaking or yielding.

2. **Stiffness (S-16) :** It is the ability of a material to resist deformation under stress.
3. **Elasticity :** It is the ability of a material to regain its original shape after deformation when the external forces are removed.
4. **Plasticity :** It is the property of a material which retains the deformation produced under load permanently. This property is useful for materials used in forgings, in stamping images on coins and in ornamental work.
5. **Ductility (S-15) :** It is the property of a material enabling it to be drawn into wire with the application of a tensile force.
    e.g. mild steel, copper, aluminium, nickel, zinc, tin and lead.
6. **Brittleness :** It is the property of breaking of a material with little permanent distortion.
    e.g. cast iron.
7. **Malleability :** It is the property of material which is rolled or hammered into thin sheets.
    e.g. lead, soft steel, wrought iron, copper and aluminium.
8. **Toughness :** It is the property of a material to resist fracture due to high impact loads like hammer blows. This property is desirable in parts subjected to shock and impact loads.
9. **Machinability :** It is the property of a material which refers to a relative case with which a material can be cut.
    e.g. Brass can be easily machined than steel.
10. **Resilence (S-16) :** It is the property of a material to absorb energy and to resist shock and impact loads.
    e.g. Spring material.
11. **Creep (S-15, 16) :** When a part is subjected to a constant stress at high temperature for a long period of time, it will undergo a slow and permanent deformation called creep.
    e.g. Designing internal combustion engines, boilers and turbines.
12. **Fatigue :** When a material is subjected to repeated stresses, it fails at stresses below the yield point stresses. Such type of failure of a material is known as fatigue.
    e.g. This property is considered while designing shafts, connecting rod, springs, gears etc.
13. **Hardness :** It is very important property of the metals and has a wide variety of meanings. It is the ability of a metal to cut another metal.

1.3.3.2 Selection of Materials

> **Q.1.** *What factors are to be considered while selection of the materials for design of machine element ?*
>
> **(S-15)**

Factors are to be considered while selection of the materials for design of machine element :
1. **Availability :** The material should be readily available in the market, in large enough quantities to meet the requirements. Cast iron and aluminium alloys are easily available in market.
2. **Material cost :** For every application there is a limiting cost beyond which designer can not afford. When this limit get exceeded, the designer consider other alternative material.
3. **Mechanical properties :** It is a technical factor governing the selection of material. They include strength under fluctuations, static load, elasticity, stiffness, toughness, hardness. Depending upon the working conditions and requirements, the properties are considered and material is selected.
    e.g. material for connecting rod should be capable to withstand fluctuating stress induced, so here endurance limit becomes the selection criteria.
4. **Manufacturing considerations :** Machinability of material is an important consideation in selection. When material is complex shaped, casting property is important. The manufacturing processes such as forging, casting, rolling, machining, extrusion etc. governs the selection of material.
5. **Manufacturing cost :** It includes cost of processing the material into finished goods.

1.4 DESIGN PROCEDURE   (W-15, 17)

> **Q.1.** *Write the design procedure of designing any machine element.*

1. Any design procedure depends on specification of an element.

2. **Mechanisms :** Selection of mechanism or group of mechanism which will give desired motion.

3. **Force analysis :** Analyse the forces which are acting on each member of machine parts.

4. **Material selection :** Select proper material which is suitable for machine parts.

5. **Design of an element :** Finding out the size of each member of the machine by considering the force acting on the member of permissible stresses for the material used. The material should not deflect or deform than permissible limit.

6. **Modification :** Modify the size of the member which is easy to manufacture.

7. **Detailed drawing :** Draw the detailed drawing of each component and assembly of the machine with complete specification for the manufacturing processes.

8. **Production :** The component, as per drawing is manufactured in the workshop.

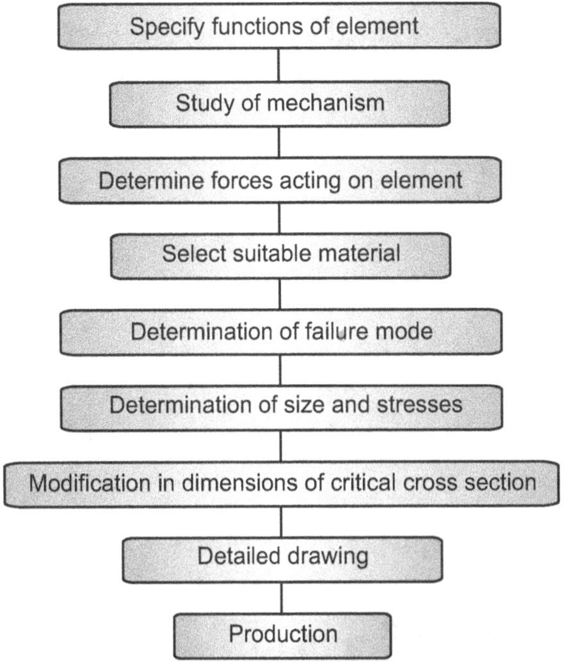

General Procedure in Machine Design

1.5 STANDARDIZATION   (W-16, 17, 18; S-17, 18)

Definition : It is defined as "obligatory norms, to which various characteristics of a product should conform". The characteristics include materials, dimensions and shape of the component, method of testing and method of marking, packing and storing of the product.

Advantages :   (S-15)

1. Interchangeability of product or element is possible.
2. Mass production is easy.
3. Rate of production increases.
4. Reduction of labour cost.
5. Limits the variety of size and shape of product.
6. Overall reduction in cost of production.
7. Improves overall performance, quality and efficiency of product.
8. Better utilization of labour, machine and time.

1.5.1 Interchangebility   (W-17)

The term interchangeability is normally employed for the mass production of identical items within the prescribed limits of sizes.

In order to maintain the sizes of the part within a close degree of accuracy, a lot of time is required. But even then there will be small variations. If the variations are within certain limits, all parts of equivalent size will be equally fit for operating in machines and mechanisms. Therefore certain variations are recognized and allowed in the sizes of the mating parts to give the required fitting.

This facilitates to select at random from a large number of parts for an assembly and results in a saving in the cost of production.

In order to control the size of finished part, with due allowance for error, for interchangeable parts is called *limit system*.

It may be noted that, when an assembly is made of two parts, the part which enters into the other, is known as *enveloped surface* (or *shaft*) and the other in which one enters is called *enveloping surface* (or *hole*).

Limits of Sizes :

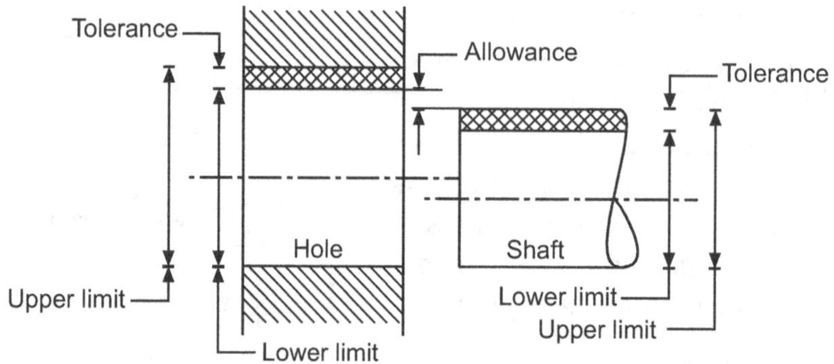

Fig. 1.1

1.6 PREFERRED NUMBERS   (S-17)

> **Q.1.** *What are preferred numbers ? How are they useful ? How are the series of preferred numbers designated ?*

In industrial design, preferred numbers are standard guidelines for choosing exact product dimensions within a given set of constraints. Product developers must choose numerous lengths, distances, diameters, volumes and other characteristic quantities.

While all of these choices are constrained by considerations of functionality, usability, compatibility, safety or cost, there usually remains considerable leeway in the exact choice for many dimensions.

1.6.1 Purposes of Preferred Numbers

1. Using the preferred numbers increases the probability of compatibility between objects designed at different times by different people.

2. They are chosen such that when a product is manufactured in many different sizes, these will end up roughly equally spaced on a logarithmic scale. They therefore, help to minimize the number of different sizes that need to be manufactured or kept in stock.

The series of preferred numbers are designated as R5, R10, R20 and R40 respectively. These four series are called *basic series*. The other series are called *derived series*.

The most basic R5 series consists of following five rounded numbers.

R5 : 1.00 1.60 2.50 4.00 6.30

e.g. (i) If our design constraints tells us that the two screws in our gadget should be placed between 32 mm and 55 mm apart, we make it 40 mm, because 4 is in the R5 series of preferred numbers.

(ii) If we want to produce a set of nails with length between roughly 15 mm and 300 mm, then the application of the R5 series would lead to a product of 16 mm, 25 mm, 40 mm, 63 mm, 100 mm, 160 mm and 250 mm long nails.

The table below shows the basic series of preferred numbers according to IS : 1076 (Part I) - 1985 (Reaffirmed 1990)

Table 1.1 : Preferred numbers

R5 series ($\phi = 1.6$)	R10 series ($\phi = 1.25$)	R20 series ($\phi = 1.12$)	R40 series ($\phi = 1.06$)
1.00	1.00	1.00	1.00
-	-	-	1.06
-	-	1.12	1.12
-	-	-	1.18
-	1.25	1.25	1.25
-	-	-	1.32
-	-	1.40	1.40
-	-	-	1.50
1.60	1.60	1.60	1.60
-	-	-	1.70
-	-	1.80	1.80
-	-	-	1.90
-	2.00	2.00	2.00
-	-	-	2.12
-	-	2.24	2.24
-	-	-	2.36
2.50	2.50	2.50	2.50
-	-	-	2.65
-	-	2.80	2.80
-	-	-	3.00
-	3.15	3.15	3.15
-	-	-	3.35
-	-	3.55	3.55
-	-	-	3.75
4.00	4.00	4.00	4.00
-	-	-	4.25
-	-	4.50	4.50
-	-	-	4.75
-	5.00	5.00	5.00
-	-	-	5.30
-	-	5.60	5.60
-	-	-	6.00
6.30	6.30	6.30	6.30
-	-	-	6.70
-	-	7.10	7.10
-	-	-	7.50
-	8.00	8.00	8.00
-	-	-	8.50
-	-	9.00	9.00
-	-	-	9.50
10.00	10.00	10.10	10.10

1.7 PRINCIPLES OF ERGONOMICS   (W-16; S-18)

Q.1. *Explain role of ergonomics in automobile design.*

Q.2. *What is ergonomics ? State its scope in machine design.*   **(W-14)**

Ergonomics is defined as "the relationship between man and machine and the application of anatomical, physiological and psychological principles to solve the problems arising from man-machine relationship". The word 'ergonomics' is taken from greek words ergon (means work) and nomos (means natural laws).

1.   However ergonomists consider man-machine joint system, forming a closed loop as shown in Fig. 1.2.

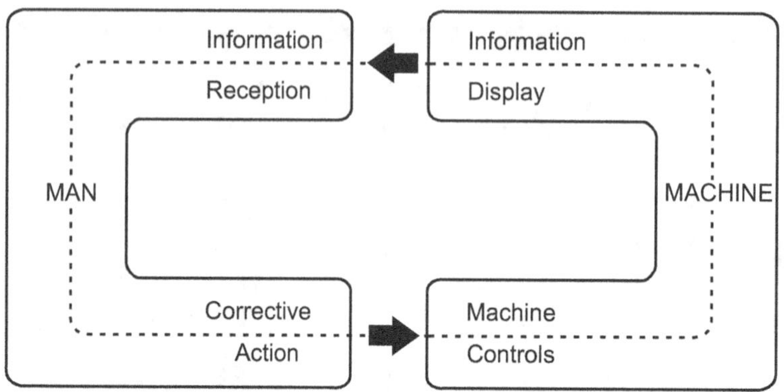

Fig. 1.2 : Man-Machine closed loop system

From display instruments, the operator gets the information about the operations of the machine. If he feels that a correction is necessary, he will operate the levers or controls. This in turn will alter the performance of the machine, which will be indicated on display panels. The contact between man and machine in this closed loop system arises at two places - display instruments, which give information to the operator, and controls with which the operator adjusts the machine.

2.   Moving scale or dial type instruments are used for quantitative measurements, while lever-type indicators are used for setting purposes.

The basic objective behind the design of displays is to minimize fatigue to operator, who has to observe them continuously.

The ergonomic considerations in design of displays are as follows :

1.   The scale on the dial indicator should be divided in suitable numerical progression like 0 – 10 – 20 – 30 and not 0 – 5 – 30 – 35.

2.   The number of subdivisions between numbered divisions should be minimum.

3.   The size of letters or numbers on the indicator should be as follows :

$$\text{Height of letter or number} \geq \frac{\text{Reading distance}}{200}$$

4.   Vertical figures should be used for stationary dials, while radially oriented figures are suitable for rotating dials.

5.   The pointer should have a knife-edge with a mirror in the dial to minimize parallax error.

The aim of ergonomics is to reduce the operational difficulties present in man-machine joint system and to thereby decrease the resulting physical and mental stress.

1.7.1 Ergonomic Considerations

Anthropometry, physiology and psychology are the components of ergonomics :

1.　Anthropometry : With the help of anthropometry, dimensions of the components are finalized, so that they can be easily handled by operator without fatigue and with consistence efficiency. For example, diameter of steering wheel, distance from chair to pedals.

2.　Physiology : With the help of physiology, components are designed to be operated by hand or first force.

Gear shifting, steering wheel are designed to be operated by hand because they require speed and accuracy which is imparted by hand, and brake pedal, clutch pedal etc. are designed to be operated by foot force because they require great amount of force than accuracy.

3.　Psychology : Psychology affects mode of operation. For e.g. size, colour and push operation of emergency stop button of any machine. The size of emergency control is made large and painted in red so that they can be easily identified and always they are push operated.

All these components make design of automobile components user friendly.

1.7.2 Importance of Ergonomic　　(W-15)

1. The importance of ergonomics is to reduce the operational difficulties present in man-machine joint system and thereby reduce the resulting physical and mental stresses.

2. Ergonomics gives exhaustive details of the dimension and resisting forces of different control elements.

3. Standardization of automobile system controls as per the regional anthro-pometry.

1.8 AESTHETICS IN DESIGNING AUTOMOBILE COMPONENTS　(S-15; W-17,18)

Q.1.　*Define the term aesthetics.*

Each product has a definite purpose. It has to perform specific functions to the satisfaction of customers. The contact between the product and the people arises due to the necessity of this functional requirement.

However, when there are a number of products in the market, having the same qualities of efficiency, durability and cost, the customer is naturally attracted towards the most appealing product. The external appearance is an important feature, which gives grace and luster to the product and dominates the market. This is true for consumer durables like automobiles, household appliances and audio-visual equipment.

There is a relationship between functional requirement and appearance of a product. In many cases, functional requirements results in shapes, which are aesthetically pleasing.

E.g. The evolution of the streamlined shape of the 'Boeing' is the result of studies in aerodynamics for effortless speed.

The selection of proper colour is an important consideration in product aesthetics. Many colours are associated with different moods and conditions. Morgan has suggested the meaning of these colours.

Following table gives Morgan's colour code :

Colour	Meaning
Red	Danger-Hazard-Hot
Orange	Possible danger
Yellow	Caution
Green	Safety
Blue	Caution-cold
Grey	Dull

The external appearance of the product also depends upon following factors :

1.   Rigidity and resilience.

2.   Tolerances and surface finish.

3.   Motion of individual components.

4.   Materials.

5.   Manufacturing methods and noise.

1.8.1 Importance of Aesthetics   (W-15)

1.   Aesthetics gives the functional requirements and appearance of the product. The functional requirements results in shapes which are aestheticaly pleasing.

2.   Aesthetics gives the cumulative effect of a number of factors like form, colour, rigidity, and tolerance, motion of individual components, manufacturing method and noise.

3.   Better surface finish always attracts the observers which increases the customers satisfaction.

Practice Questions

1.   What is meant by machine design ?

2.   Explain the basic procedure of machine design.

3.   Name three basic types of failure of the machine elements ?

4.   Give reasons for the following statements :

    (i)   Flywheel is made of cast iron.

    (ii)   Automobile body is made of low carbon steel.

    (iii)   Shafts are made of steel.

    (iv)   Bolts are made of free cutting steels.

    (v)   Crank shaft is made of alloy steels.

5.   What is standardization ? What are its advantages ?   (S-14)

6.   Define standardization.

7. What is relationship between the functional requirement and the appearance of a product ?

8. Discuss the ergonomic considerations in design of displays and controls.

9. What are preferred numbers ?

10. What is meant by bolts of uniform strength ? Give examples.

MSBTE Questions and Answers (As per G-Scheme)

Summer 2016

1. Describe fatigue and endurance limit. **(4 M)**

Ans. Refer Article 1.2.

2. Define : (i) Strength, (ii) Stiffness, (iii) Creep, (iv) Resilence. **(4 M)**

Ans. Refer Article 1.3.1.

3. Enlist types of failure and describe any three of them. **(4 M)**

Ans. Refer Article 1.2.

Winter 2016

1. Explain design consideration in automobile design. **(4 M)**

Ans. Refer Article 1.3.

2. Explain Ergonomic aspects of machine design. **(4 M)**

Ans. Refer Article 1.7.

3. Define standardization and state the four advantages of it. **(4 M)**

Ans. Refer Article 1.5.

Summer 2017

1. Give classification of design. **(4 M)**

Ans. Refer Article 1.1.1.

2. Explain the following terms : (i) Concept of standardization, (ii) Preferred number. **(6 M)**

Ans. Refer Articles 1.5, 1.6.

Winter 2017

1. Explain aesthetic considerations in designing of automobile components. **(4 M)**

Ans. Refer Article 1.8.

2. Explain the term : (i) Standardisation and (ii) Interchangeability in design. **(4 M)**

Ans. Refer Articles 1.5. and 1.5.1.

2. State the general design considerations. **(4 M)**

Ans. Refer Article 1.3.

Summer 2018

1. Explain Ergonomic aspects of machine design.   **(4 M)**

Ans. Refer Article 1.7.

2. Explain any eight design considerations in machine design.   **(4 M)**

Ans. Refer Article 1.3.

3. Explain the term standardization. State any four advantages of it.   **(4 M)**

Ans. Refer Article 1.5.

Winter 2018

1. State the eight considerations in machine design.   **(4 M)**

Ans. Refer Article 1.3.

2. Define standardization and state the four advantages of it.   **(4 M)**

Ans. Refer Article 1.5.

3. Explain aesthetic consideration in designing automobile components.   **(4 M)**

Ans. Refer Article 1.8.

Chapter 2

STRESSES IN AUTOMOBILE COMPONENTS

Weightage of Marks = 12, Teaching Hours = 12

Syllabus

2.1   Normal and Shear Stresses, Crushing Stress, Bearing Pressure, Torsional and Bending Stresses, Principal Stresses, Variable Stresses, Impact Stresses, Resilience

2.2   Stress-strain Diagram and its Uses

2.3   Working Stress, Factor of Safety, Selections of FOS

2.4   Theories of Failure Under Static Loading - Maximum Principal or Normal Stress Theory, Maximum Shear Stress Theory, Maximum Distortion Energy Theory

2.5   Fatigue, Endurance Limit, FOS for Fatigue Loading, S-N Curve

2.6   Stress Concentration, Causes and Remedies

2.7   Load Factor, Service Factor and Their Applications

About this Chapter

After reading this chapter, students will be able to :

- Identify stress(s) induced in the given component for the given load condition with justification.
- Describe mechanical properties/theories of failure of the given component material.
- Use safety factor in calculating dimensions of the given component under given load conditions.
- Suggest suitable remedies to reduce stress concentration for the given component/s with justification.

2.1 STRESS ANALYSIS

2.1.1 Types of External Loads

There are two types of external load acting on the component (a) static and (b) dynamic.

(a) Static load : Static load is defined as "a load, which do not vary in magnitude or direction with respect to time, after it has been applied".

e.g.   1. Dead weights of machinery,

2. Load acting on the walls,

3. Cover of a pressure vessel subjected to constant pressure.

(b) Dynamic load : Dynamic load is a load which vary in magnitude or in direction or in both, magnitude and direction with respect to time, after it has been applied. There are two types of dynamic load :

(i)   Cyclic load,

(ii) Impact load.

(i) Cyclic load (W-15) : It is a load which when applied, varies in magnitude in a repetitive cycle manner, from tension to compression or oscillating about some mean value.

Cyclic loading is "the application of repeated fluctuating stresses, strains, or stress intensities to locations on structural components".

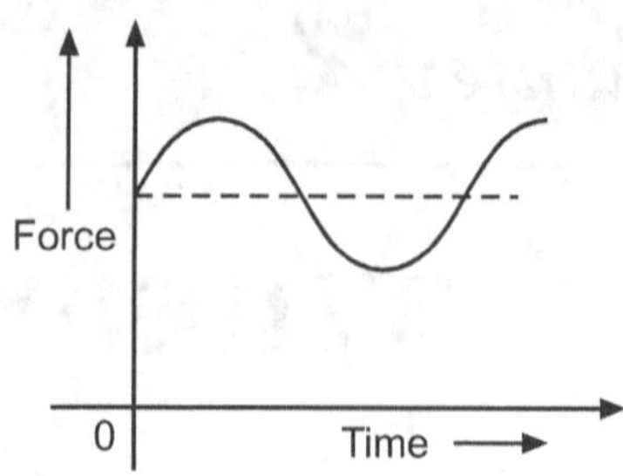

Fig. 2.1

(ii) Impact load : It is a load which is applied suddenly to a member usually at a high velocity. Due to damping in the material, the excess of energy transferred during impact is converted into heat and the load in time reaches to its static value.

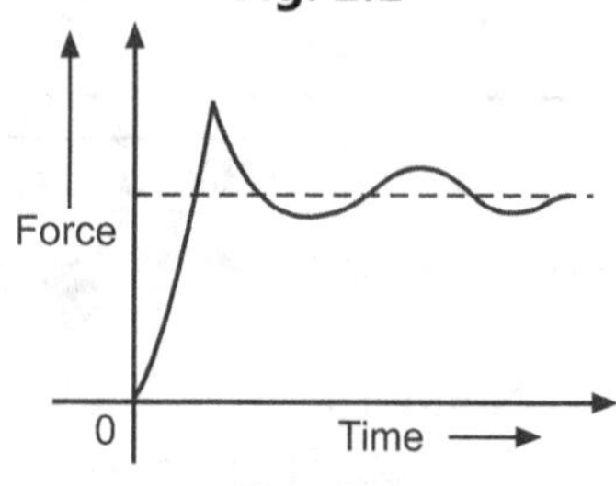

Fig. 2.2

2.1.2 Types of Induced Stresses

Stress : When some external system of forces or loads act on a body, the internal forces are setup at various sections of the body, which resist the external forces. This internal force per unit area at any section of the body is known as *unit stress* or simply *stress*. It is denoted by 'σ' (sigma).

$\therefore$

$$\sigma = \frac{P}{A}$$

where, P = Force or load acting on a body

and A = Cross-sectional area of the body

S.I. unit of stress is Pascal (P_a).

$\therefore$ $1P_a = 1 \text{ N/m}^2$

Also, $1 \text{ MP}_a = 1 \times 10^6 \text{ N/m}^2 = 1 \text{ N/mm}^2$

$1 \text{ GP}_a = 1 \times 10^9 \text{ N/m}^2 = 1 \text{ kN/mm}^2$

Types of Stresses :

1. Tensile stress : When a body is subjected to two equal and opposite axial pulls 'P' as shown in Fig. 2.3 (a), then the stress induced at any section of the body is known as *tensile stress* as shown in Fig. 2.3 (b).

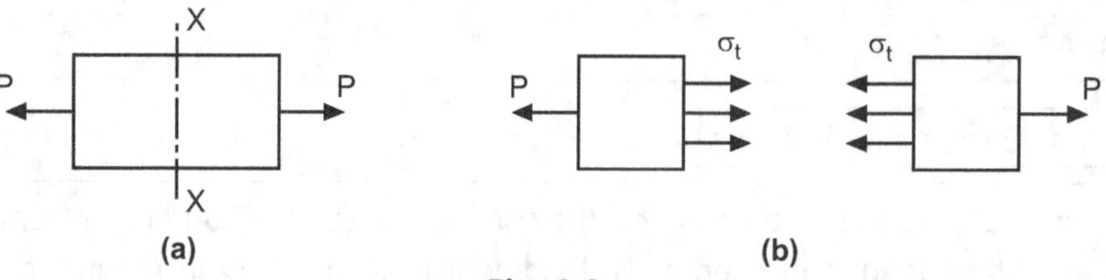

Fig. 2.3

2. Compressive stress : When a body is subjected to two equal and opposite axial push forces 'P' as shown in Fig. 2.4 (a), then stress induced at any section of the body is known as *compressive stress*, as shown in Fig. 2.4 (b).

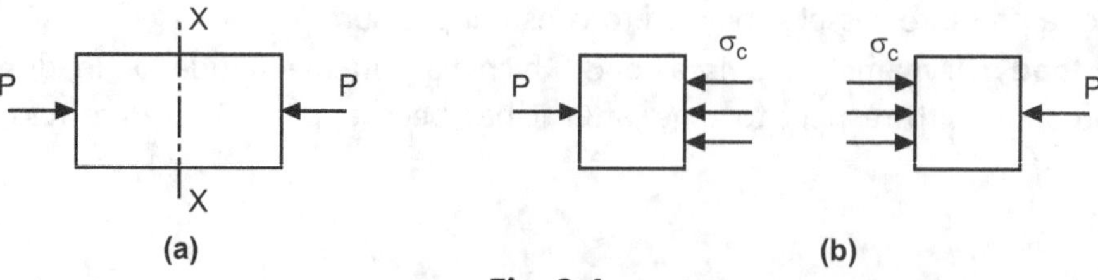

Fig. 2.4

Let,　　　　　　　　　　P　=　Axial compressive force

　　　　　　　　　　　　　A　=　Cross-sectional area

　　　　　　　　　　　　　l　=　Original length

and　　　　　　　　　　δl　=　Decrease in length

∴　Compressive stress,　$\boxed{\sigma_c \;=\; \dfrac{P}{A}}$

3.　Shear stress : When body is subjected to two equal and opposite forces acting tangentially across the resisting section, as a result of which the body tends to shear off the section, then the stress induced is called *shear stress*. Shear stress is denoted by 'τ'.

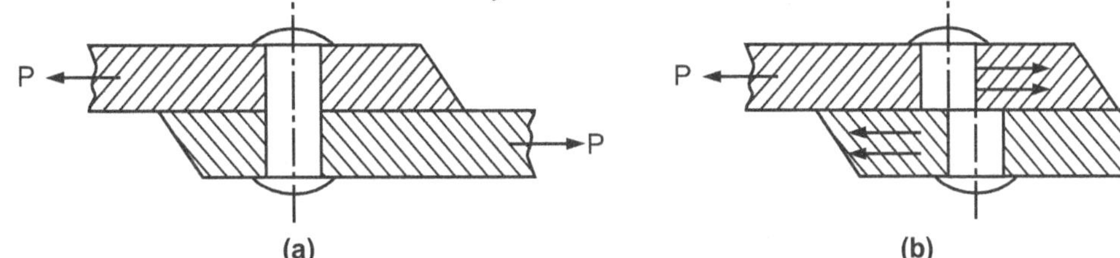

(a)　　　　　　　　　　　　　　　　　　　　(b)

Fig. 2.5

∴　　　　　Shear stress 'τ'　$=\;\dfrac{\text{Tangential force}}{\text{Resisting area}}$

For a single rivet section,

The area resisting the shear off the rivet, $A = \dfrac{\pi}{4}d^2$

and shear stress on the rivet cross-section,

$$\boxed{\tau \;=\; \dfrac{P}{A} \;=\; \dfrac{P}{\dfrac{\pi}{4}d^2} \;=\; \dfrac{4P}{\pi d^2}}$$

4.　Crushing stress or Bearing stress : The compressive stress is available between two mating members of a machine parts, which are at rest, is known as *bearing stress* or *crushing stress*.

This stress is applicable for riveted joints, cotter joints, knuckle joints etc.

(a)　Bearing stress in a riveted joint :

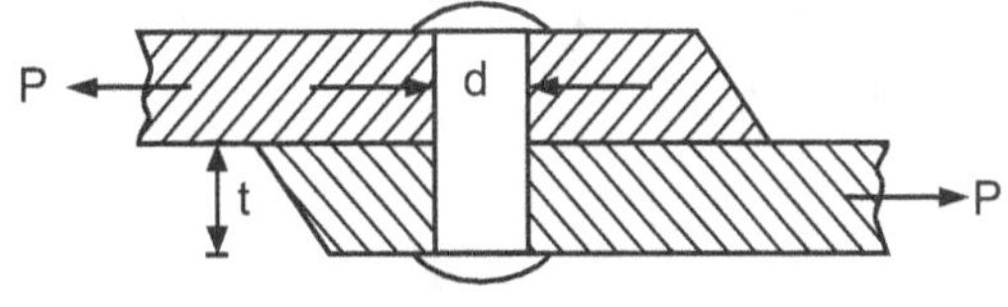

Fig. 2.6

(b)　Journal supported in a bearing :

(i)　　　　　　　　　　　　　　　　　　　　**(ii)**

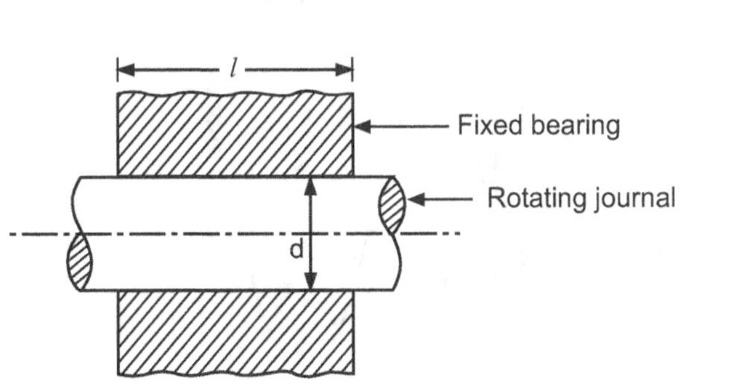

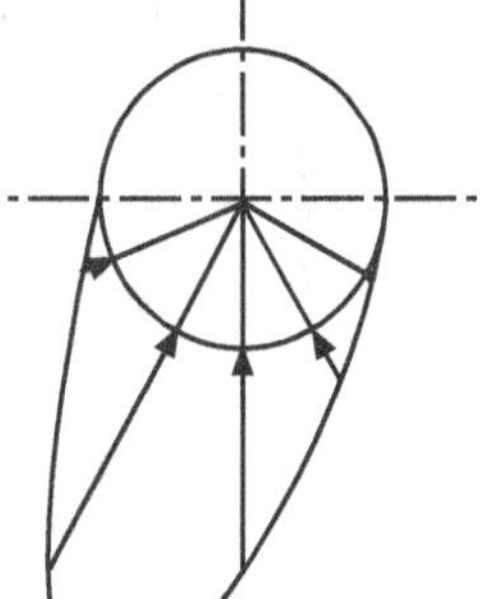

Fig. 2.7　　　　　　　　　**Fig. 2.8 : Distribution of bearing pressure**

The crushing or bearing stress is calculated by,

$$\sigma_b \text{ or } \sigma_c = \frac{P}{d.t.n}$$

where,
d = diameter of the rivet
t = thickness of the plate
d.t = projected area of the rivet
and n = number of rivets per pitch length in bearing or crushing

5. **Bending stress :** A straight beam subjected to a bending moment M_b is as shown in Fig. 2.9.

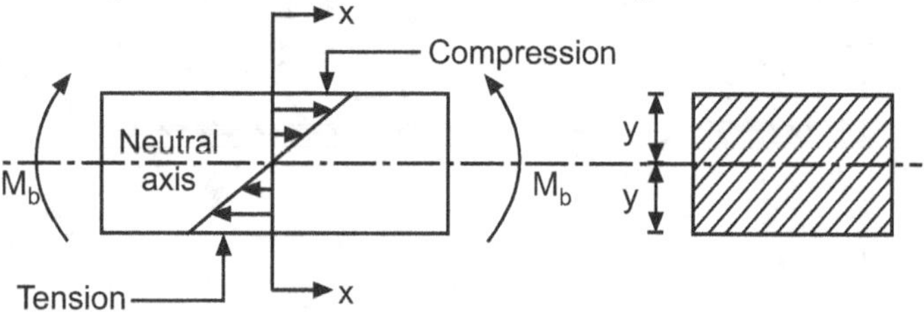

Fig. 2.9

Distribution of Bending Stress :

The beam is subjected to a combination of tensile stress on one side of the neutral axis and compressive stress on the other.

The outside fibres are in tension, while the inside fibres are in compression.

The bending stress at any fibre is given by,

$$\sigma_b = \frac{M_b Y}{I}$$

where,
σ_b = Bending stress at a distance y from neutral axis (N/mm^2)
M_b = Bending moment (N-mm)
I = Moment of inertia of the cross-section about neutral axis (mm^4)

6. **Cyclic Stresses :** (W-15)

(i) Completely reversed stresses : The stresses which vary from a minimum value to a maximum value of the opposite nature and completely reversed cycle is known as completely reversed stresses. [Fig. 2.10]

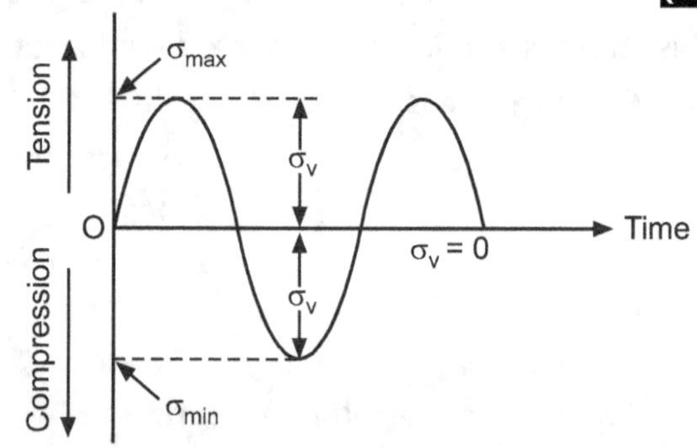

Fig. 2.10 : Completely reversed stress

(ii) Repeated stresses : The stresses which vary from zero to a certain maximum value are called repeated stresses. [Fig. 2.11]

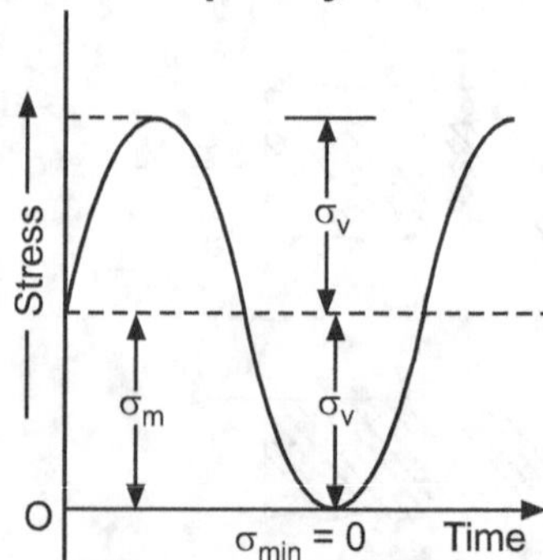

Fig. 2.11 : Repeated stress

(iii) Fluctuating stresses : The stresses which vary from a minimum value to a maximum value of the same nature are called repeated stresses. [Fig. 2.12]

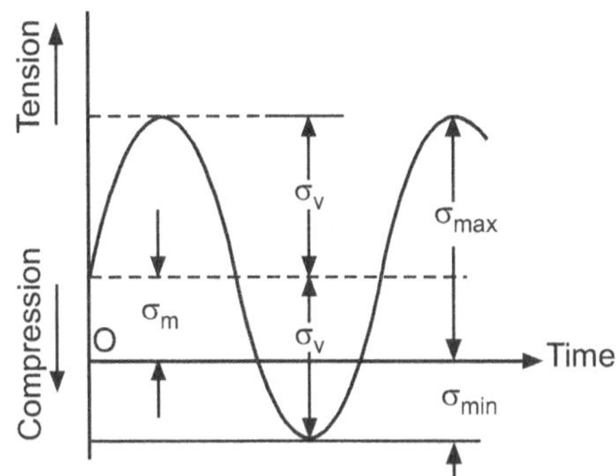

Fig. 2.12 : Fluctuating stress

7.　**Torsional stress :** A transmission shaft, subjected to an external torque is as shown in Fig. 2.13 (a).

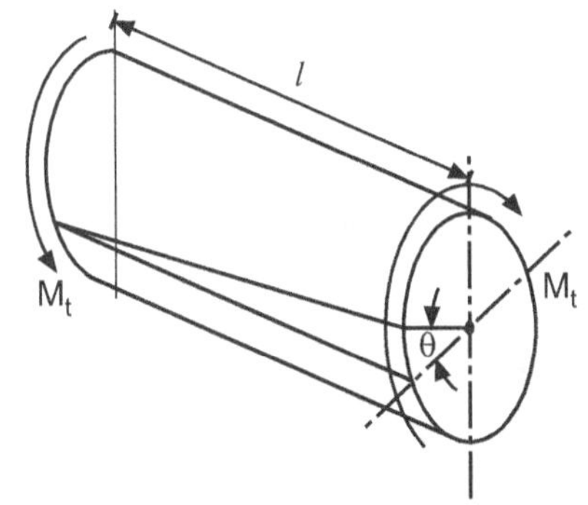

(a) Shaft subjected to torsional moment 'M$_t$'

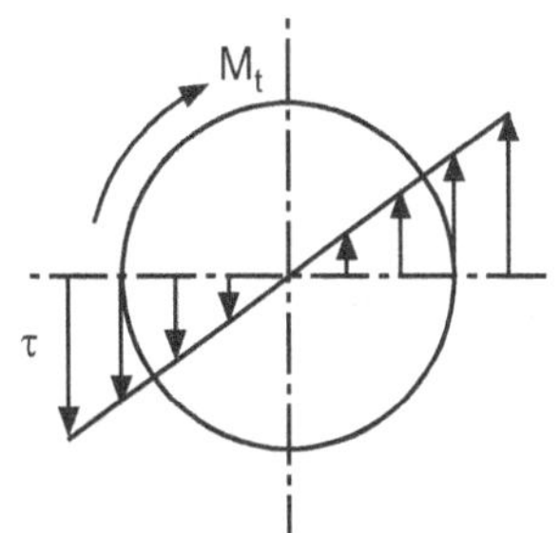

(b) Distribution of torsional shear stress

Fig. 2.13

The internal stresses, which are induced to resist the action of twist, are called *torsional shear stress*. The torsional shear stress is given by,

$$\tau = \frac{M_t\, r}{J}$$

where,　　　　　τ = Torsional shear stress at the fibre (N/mm^2)

M_t = Applied torque (N/mm)

r = Radial distance of the fibre from the axis of rotation (mm)

J = Polar moment of inertia of the cross-section about the axis of rotation (mm^4)

The distribution of torsional shear stress is shown in Fig. 2.13 (b).

The stress is maximum at the outer fibre and zero at the axis of rotation, the angle of twist is given by,

$$\theta = \frac{M_t \cdot l}{JG}$$

where,　　　　　θ = Angle of twist (radians)

l = Length of the shaft (mm)

7.　**Thermal stresses :** Whenever a body is subjected to a rise or fall in temperature, it experiences expansion or contraction. If the body is allowed to expand or contract freely, due to the change in temperature, stresses are not induced in the body. But, if the deformation of the body is prevented; stresses known as thermal stresses are induced.

Let,　　　　　l = Original length of the part, mm

t = Change in temperature, °C

α = Coefficient of thermal expansion

Then, increase in length, $\boxed{\delta l = l\,\alpha\,t}$

If the ends of a body are fixed, then preventing the expansion, thermal stresses, compressive in nature will be induced, which is given by,

$$\boxed{\sigma = E\,\alpha\,t}$$

8. Principal stresses : There are many components, which are subjected to several types of load. A transmission shaft is subjected to bending as well as torsional moment at the same time. In design, it is necessary to determine the state of stresses under these conditions.

Following Fig. 2.14 shows, a plate is subjected to two dimensional stresses. The stresses are classified into two groups :

   (i)   Normal stresses (σ_x, σ_y and σ_z).

   (ii)   Shear stresses (τ_{xy}, τ_{yz}).

The normal stress is perpendicular to the area under consideration, while the shear stress acts over the area.

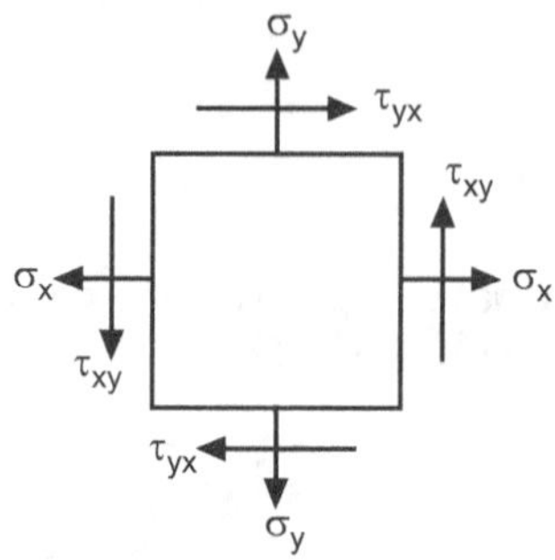

Fig. 1.14

Tensile stresses are considered as positive and compressive stresses are considered as negative. The shear stresses are positive if they act in the positive direction of the reference axis. It can be proved that,

$$\boxed{\tau_{xy} = \tau_{yx}} \qquad \ldots (2.1)$$

Following Fig. 2.15 shows the stresses acting on an oblique plane. The normal to the plane makes an angle θ with the X-axis. σ and τ are normal and shear stresses associated with this plane.

For equilibrium of forces, stresses on oblique plane are

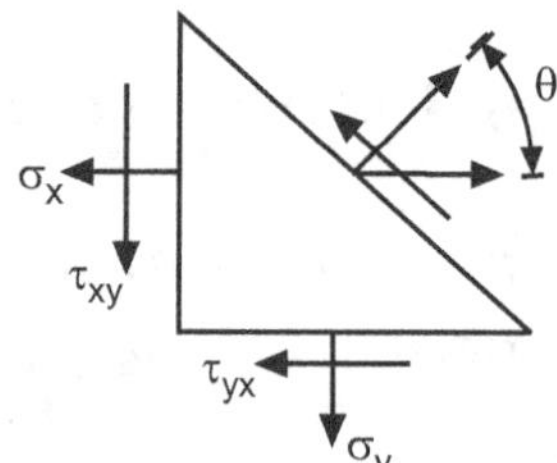

$$\sigma = \left(\frac{\sigma_x + \sigma_y}{2}\right) + \left(\frac{\sigma_x - \sigma_y}{2}\right)\cos 2\theta + \tau_{xy}\sin 2\theta \qquad \ldots (2.2)$$

Fig. 2.15

and $\qquad \tau = -\left(\dfrac{\sigma_x - \sigma_y}{2}\right)\sin 2\theta + \tau_{xy}\cos 2\theta \qquad \ldots (2.3)$

Differentiating equation w.r.t. θ and setting the result to zero, we have

$$\tan 2\theta = \frac{2\tau_{xy}}{\sigma_x - \sigma_y} \qquad \ldots (2.4)$$

Equation (2.4) defines two values of (2θ), one giving the maximum value of normal stress and other the minimum value.

If σ_1 and σ_2 are the maximum and minimum values of normal stress, then substituting equation (2.4) in equation (2.2), we get

$$\sigma_1 = \left(\frac{\sigma_x + \sigma_y}{2}\right) + \sqrt{\left(\frac{\sigma_x - \sigma_y}{2}\right)^2 + (\tau_{xy})^2} \qquad \ldots (2.5)$$

and $\qquad \sigma_2 = \left(\dfrac{\sigma_x + \sigma_y}{2}\right) - \sqrt{\left(\dfrac{\sigma_x - \sigma_y}{2}\right)^2 + (\tau_{xy})^2} \qquad \ldots (2.6)$

σ_1 and σ_2 are called *principal stresses*.

Similarity equation (2.3) is with respect to θ and the result is equated to zero, giving the following condition

$$\tan 2\theta = -\left(\frac{\sigma_x - \sigma_y}{2\tau_{xy}}\right) \qquad \ldots (2.7)$$

Substituting equation (2.7) in equation (2.3) we get,

$$\tau_{max} = \pm \sqrt{\left(\frac{\sigma_x - \sigma_y}{2}\right)^2 + (\tau_{xy})^2}$$

∴　　τ_{max} is called the *principal shear stress*.

2.2 STRESS-STRAIN DIAGRAM

Q.1.　*Draw stress-strain diagram for ductile material and explain.*　**(S-15)**

Q.2.　*Distinguish between stress-strain diagram for ductile and brittle materials.*

Q.3.　*Explain with sketch stress-strain diagram for brittle material.*

Importance : The behaviour of material and its usefulness for engineering applications can be obtained by making tensile test and plotting a curve showing variations of stress with respect to strain.

(A) Stress-Strain Diagram for Ductile Materials :　　**(W-16, 18; S-18)**

Following properties of materials can be obtained from this diagram :

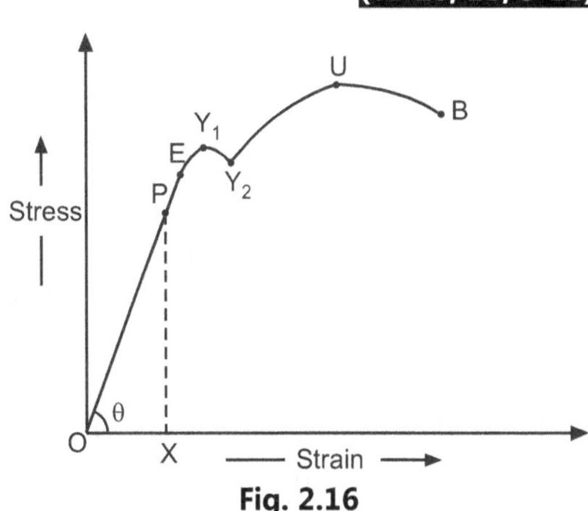

Fig. 2.16

1.　Proportional limit (Point P) : From graph, it is observed that, the stress is directly proportional to the strain upto point P, from point O and obey's Hooks law.

The term *proportional limit* is defined as "the stress at which the stress-strain curve begins to deviate from the straight line". The point P indicates the proportional limit.

2.　Modulus of elasticity (Point E) : The *modulus of elasticity* or *Young's modulus* (E) is "the ratio of stress to strain upto point P". It is given by the slope of line OP.

Therefore,　　　$E = \tan\theta = \dfrac{PX}{OX} = \dfrac{Stress}{Strain}$

3.　Elastic limit (Point E) : Even if the specimen is stressed beyond point P and upto point E, it will regain its initial size and shape when the load is removed. This indicates that the material is in elastic stage upto point E.

Therefore, E is called elastic limit. This is the maximum stress which the material can bear without any permanent deformation.

4.　Yield strength (Points Y₁ and Y₂) : When the specimen is stressed beyond point E, plastic deformation occurs and material starts yielding. During this stage, it is not possible to recover the initial size and shape of the specimen on the removal of the load.

Beyond point 'E', the strain increases at a faster rate upto point Y_1 without any increase in stress.

In case of mild steel, it is observed that there is small reduction in load and the curve drops down to point Y_2 immediately after yielding starts.

The points Y_1 and Y_2 are called upper and lower yield points respectively.

5.　Ultimate tensile strength (Point U) : After the yield point Y_2, plastic deformation of the specimen increases. The material becomes stronger due to strain hardening and higher and higher load is required to deform the material. Finally, the load and corresponding stress reaches a maximum value, as given by point U. The stress corresponding to point U is called ultimate strength. The *ultimate strength* is "the maximum stress that can be reached in the tension test".

The diameter of the specimen begins to decrease rapidly beyond maximum load point U. There is localized reduction in cross-sectional area called **necking.**

6.　Breaking point (Point B) : As test progresses, the cross-sectional area at the neck decreases rapidly and fracture takes place at the narrowest cross-section of the neck. This fracture is shown by point 'B' on the diagram. The stress at the time of fracture is called *breaking strength*.

7.   **Percentage elongation :** After the fracture, the total length of the specimen is measured.

The *percentage elongation* is defined as "the ratio of the increase in the length of the gauge section of the specimen to its original length", expressed in per cent.

$$\therefore \quad \boxed{\text{Percentage elongation} \ = \ \left(\frac{l - l_o}{l_o}\right) \times 100}$$

where, l = Gauge length

             l_o = Original length

Ductility is measured by a percentage elongation.

8.   **Percentage reduction in area :** *Percentage reduction in area* is defined as "the ratio of decrease in cross-sectional area of the specimen after fracture to the original cross-sectional area, expressed in per cent".

$$\therefore \quad \boxed{\% \ \text{reduction in area} \ = \ \left(\frac{A_o - A}{A_o}\right) \times 100}$$

where, A_o = Original cross-sectional area of the test specimen

           A = Final cross-sectional area after fracture

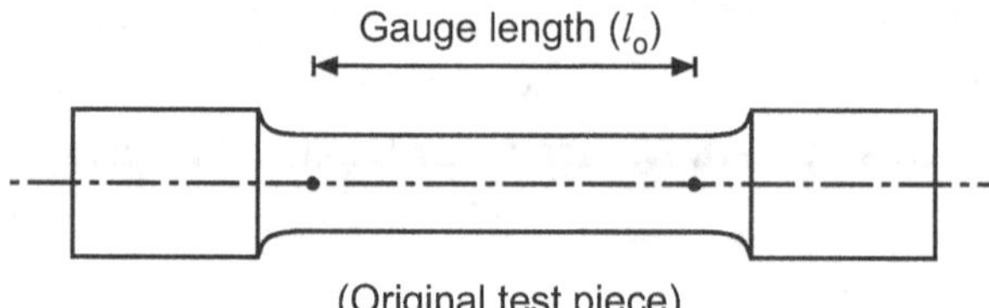

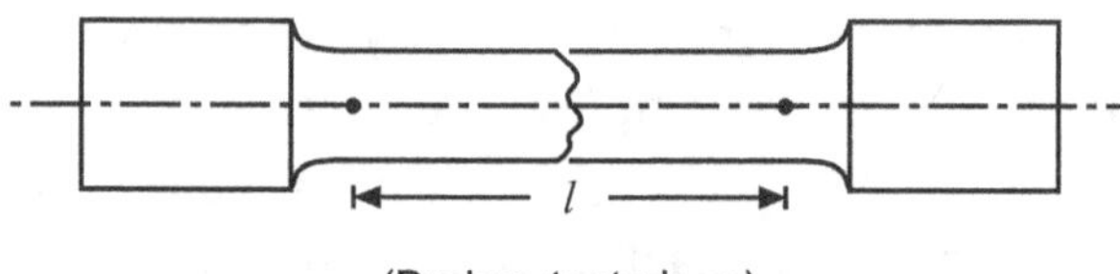

Fig. 2.17

(B) Stress-Strain Diagram for Brittle Material

The stress-strain diagram for brittle material like cast iron is shown in Fig. 2.18. It is observed that such materials do not exhibit the yield point. The deviation of stress-strain curve from straight line begins very early and fracture occurs suddenly at point U with very small plastic deformation and without necking. Therefore, ultimate tensile strength is considered as failure criterion in brittle materials.

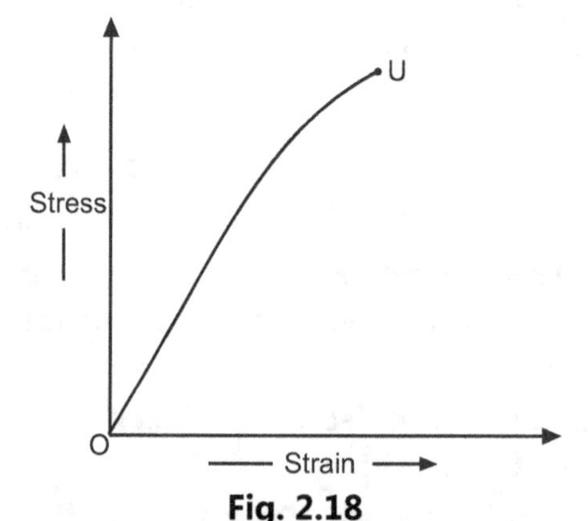

Fig. 2.18

2.3 WORKING STRESS

While designing machine parts, it is desirable to keep the stress lower than the maximum or ultimate stress at which failure of material takes place. This stress is known as *working stress* or *design stress*. It is also known as *safe* or *allowable stress*.

2.3.1 Types of Working Stress

> **Q.1.** *Define factor of safety for ductile material and state the four factors considered while selecting factor of safety.*
>
> **Q.2.** *What is factor of safety ? How and why it is selected ?*

1. Static Load :

The following relations for factor of safety are for static loading.

Factor of Safety :   **(W-17; S-17, 18)**

In general, factor of safety is defined as "the ratio of the maximum stress to the working stress".

$$\text{Mathematically, Factor of safety} = \frac{\text{Maximum stress}}{\text{Working or Design stress}}$$

For ductile materials : For example, mild steel, where yield point is clearly defined, the factor of safety is based upon the yield point stress.

$$\therefore \quad \text{Factor of safety} = \frac{\text{Yield point stress}}{\text{Working or Design stress}}$$

For brittle materials : For example, cast iron, where yield point is not clearly defined,

$$\therefore \quad \text{Factor of safety} = \frac{\text{Ultimate stress}}{\text{Working or Design stress}}$$

2. Variable or Fatigue Load :

When a component is subjected to fatigue loading, the endurance limit is the criterion for failure. Therefore, the factor of safety should be based on endurance limit.

$$\therefore \quad \text{Factor of safety, (F.O.S)} = \frac{\text{Endurance limit stress}}{\text{Design or working stress}} = \frac{\sigma_e}{\sigma_d}$$

where,	σ_e = Endurance limit stress for completely reversed stress cycle

and	σ_y = Yield point stress

For steel,	σ_e = 0.8 to 0.9 σ_y

## 2.3.2 Selection of Factor of Safety	(S-15, 18; W-17, 18)

Q.1. *List the important factors that influence the magnitude of factor of safety.*

The selection of factor of safety to be used in designing any machine component depends upon a number of considerations, which are as follows :

1. The reliability of applied load.
2. The certainty as to exact mode of failure.
3. The machine component is working in corrosive atmosphere.
4. The extent of simplifying assumptions.
5. The extent of localized stresses.
6. The extent of initial stresses setup during manufacture.
7. The extent of loss of life if failure occurs.
8. The extent of loss of property if failure occurs.

2.4 THEORIES OF FAILURE

Q.1. *Explain maximum shear stress theory of failure with example.*

Q.2. *Explain following theories of failure :*

(a) Maximum principal stress theory.

(b) Maximum shear stress theory.

The principal theories of failure for a member subjected to bi-axial stress under static loading are as follows :

1. Maximum Principal Stress (or Normal) Theory (Rankine's Theory):	(S-17, 18; W-16)

According to this theory, the failure or yielding occurs at a point in a member when the maximum principal or normal stress in a bi-axial stress system reaches the limiting strength of the material in a simple tension test.

Since, the limiting strength for ductile materials is yield point stress and for brittle materials the limiting strength is ultimate stress, therefore according to the above theory, taking Factor Of Safety (F.O.S.) into consideration, the maximum principal or normal stress in a bi-axial stress system is given by,

$$\sigma_{t_1} = \frac{\sigma_{yt}}{\text{F.S.}} \text{, for ductile materials}$$

$$\sigma_{t_1} = \frac{\sigma_u}{F.S.} \text{, for brittle materials}$$

where,　　　　　　　σ_{yt} = Yield point stress in tension as determined from simple tension test

and　　　　　　　　σ_u = Ultimate stress

　　　　　　　　　　σ_{t_1} = Maximum principal stress

2. Maximum Shear Stress Theory (Guest's or Tresca's Theory) :　　　(W-17)

According to this theory, the failure or yielding occurs at a point in a member when the maximum shear stress in a bi-axial stress system reaches a value equal to the shear stress at yield point in a simple tension test.

Mathematically,　　　$$\boxed{\tau_{max} = \frac{\sigma_{yt}}{F.S.}}$$　　　　　... (2.8)

where,　　　　　　　　τ_{max} = Maximum shear stress in a bi-axial stress system

　　　　　　　　　　　σ_{yt} = Shear stress at yield point, as determined from simple tension test,

and　　　　　　　　　F.S. = Factor of Safety

Since, the shear stress at yield point in a simple tension test is equal to one-half the yield stress in tension, therefore the equation (2.8) may be written as,

$$\boxed{\tau_{max} = \frac{\sigma_{yt}}{2 \times F.S.}}$$

This theory is mostly used for designing members of ductile materials.

3. Distortion Energy Theory (Hencky and Von Mises Theory) :

According to this theory, the failure or yielding occurs at a point in a member when the distortion strain energy (also called shear strain energy) per unit volume in a bi-axial stress system reaches the limiting distortion energy per unit volume as determined from the simple tension test.

Mathematically,　$$(\sigma_{t_1})^2 + (\sigma_{t_2})^2 - 2\sigma_{t_1} \times \sigma_{t_2} = \left(\frac{\sigma_{yt}}{F.S.}\right)^2$$

This theory is mostly used for ductile materials.

2.5 VARIABLE STRESSES IN MACHINE PARTS

Q.1.　*Define the terms : (i) Fatigue and (ii) Endurance limit with suitable example.*　　(S-14)

Q.2.　*Define variable stresses in machine parts. Describe completely reverse bending cycle.*

Q.3.　*Describe the concept of fatigue and endurance limit.*　　(S-14)

Fatigue (S-16; W-16, 17, 18) : When a material is subjected to repeated stresses, it fails at stresses below the yield point stress, such type of failure of a material is known as *fatigue*.

The failure is caused by means of a progressive crack formation which are usually fine and of microscopic size. The failure may occur even without any prior indication. The fatigue of material is effected by the size of the component, relative magnitude of static and fluctuating loads, and the number of load reversals.

Endurance limit (S-16, 17; W-16, 17, 18) : It is denoted by 'σ_e'. It is defined as "maximum value of the completely reversed bending stress which a polished specimen can withstand without failure, for finite number of cycles".

The term endurance limit is used for reversed bending only.

Endurance strength : It is defined as "the safe, maximum stress which can be applied to the machine part working under actual conditions".

2.5.1 S-N Diagram for Variable Stresses   (W-18, S-17)

R. R. Moore Rotating Beam Fatigue Machine :

A schematic diagram of rotating beam fatigue testing machine is shown in Fig. 2.19. The specimen acts as 'rotating beam' subjected to bending moment. Therefore, it is subjected to completely reversed stress cycle. Changing the bending moment by addition or subtraction of weights can vary the stress amplitude. The specimen is rotated by electric motor and the number of revolutions, before the appearance of the first fatigue crack is recorded, on a revolution counter.

In each test, two readings are taken i.e. stress amplitude (S_f) and number of stress cycles (N). These readings are used as two co-ordinates of a point, called failure point on S-N diagram. To determine the endurance limit of a material, a large number of tests are to be carried out.

The results of these tests are plotted by means of a S-N curve. The S-N curve is the graphical representation of stress amplitude (S_f) versus the number of stress cycles (N) before the fatigue failure on a log-log graph paper.

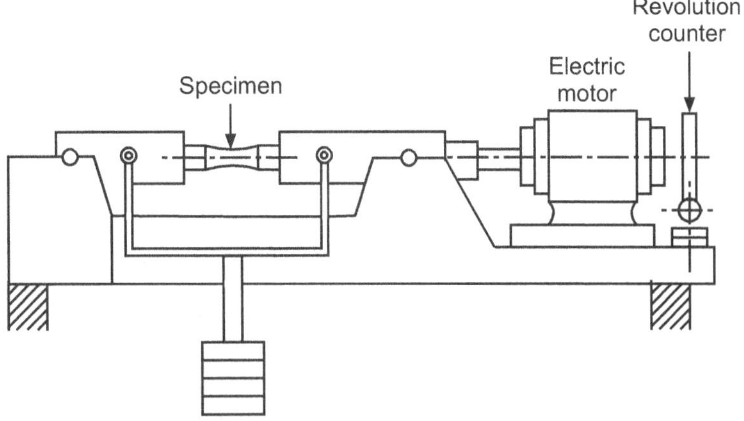

Fig. 2.19 : A rotating beam fatigue testing machine

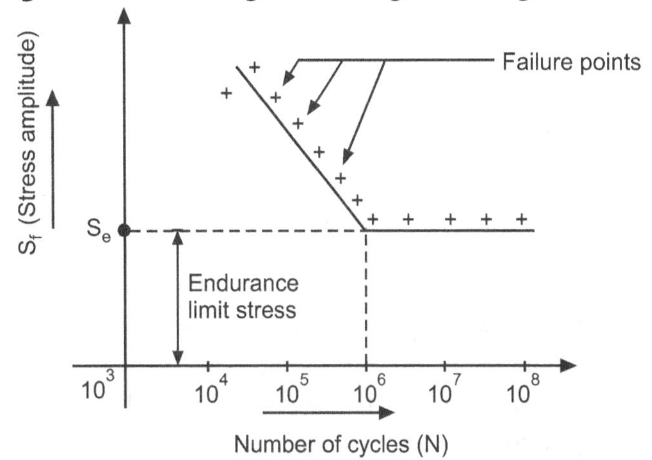

Fig. 2.20 : S-N Curve

2.6 STRESS CONCENTRATION   (W-14, 17; S-15)

Q.1. *What is stress concentration ? Explain the causes and remedies with neat sketches. State its uses.*

"The irregularity in the stress distribution caused by abrupt changes of form or shape" is called *stress concentration*.

OR

Stress concentration is defined as "the localization of high stresses due to irregularities or abrupt changes of cross-section".

It occurs at all types of stresses in the presence of holes, keyways, notches, fillets, splines, surface roughness or scratches etc.

Example 1 : Consider a member with different cross-sections under a tensile load as shown in Fig. 2.21.

The nominal stress at end A and at end B will be uniform but where cross-section is changing, a redistribution of the force within the member must take place. The simple stress distribution is no longer holds good at changed cross-section i.e. at middle region of the cross-section. Therefore, the material is stressed higher near the edge than the average value.

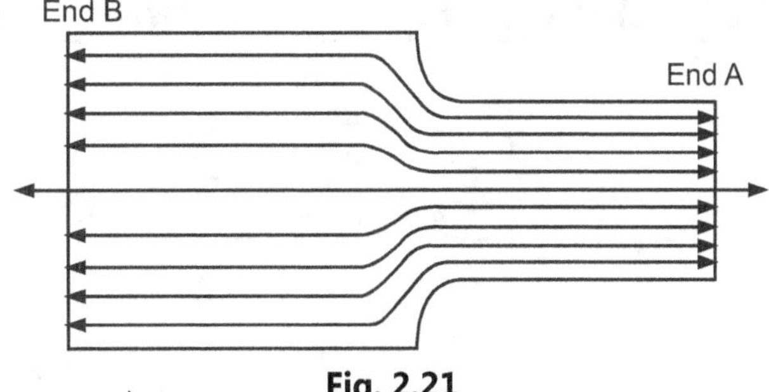

Fig. 2.21

Example 2 : A plate with a small circular hole subjected to tensile stress as shown in Fig. 2.22.

The stress distribution near the hole can be observed by keeping a model of the plate made of epoxy resin in circular polariscope. It is observed from the nature of stress distribution at the section passing through the hole that, there is sudden rise in the magnitude of stresses in vicinity of the hole.

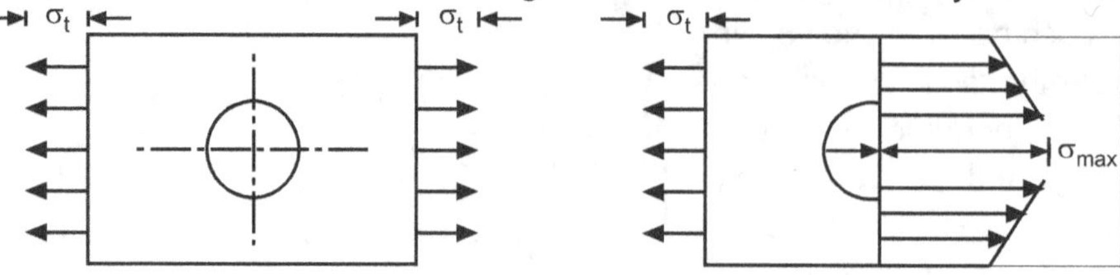

Fig. 2.22

2.6.1 Stress Concentration Factor

It is used to find out localized stresses. It is denoted by 'K_t'.

$$\therefore \qquad K_t = \frac{\text{Highest value of actual stress near discontinuity}}{\text{Nominal stress for minimum cross-section}}$$

$$\therefore \qquad K_t = \frac{\sigma_{max}}{\sigma_0} = \frac{\tau_{max}}{\tau_0}$$

where, σ_0 and τ_0 are determined by elementary equations and σ_{max} and τ_{max} are localized stresses at discontinuities.

2.6.2 Causes of Stress Concentration　　　　　(W-14, S-15)

Following are the causes of stress concentration :

1.　Variations in properties of materials : In design of machine components it is assumed that the material is homogeneous throughout the component. But it is not true for all the components. Therefore, material properties may vary from one to another due to internal cracks, flaws, cavities in welds, airholes in steel components and non-metallic foreign inclusions.

2.　Load applications : The machine components are subjected to forces. These forces act either at a point or over a small area on the component :

e.g.　(a) contact between cam and follower.

　　　(b) contact between the rail and wheel.

　　　(c) contact between the crane hook and chain.

3.　Abrupt changes in section : In order to mount gears, sprockets, pulleys and ball bearings on transmission shaft, steps are cut on the shaft and shoulders are provided from assembly considerations.

4.　Discontinuities in the component : Certain features of machine components such as oil holes or oil grooves, keyways and splines and screw threads cause irregular distribution of stress.

5.　Machining scratches : Machining scratches, stamp mark or inspection mark are surface irregularities which cause stress concentration.

2.6.3 Methods of Reducing Stress Concentration　　　　(W-14, 17; S-15)

1.

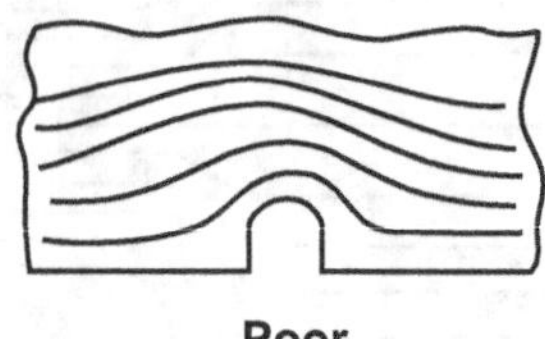

Poor

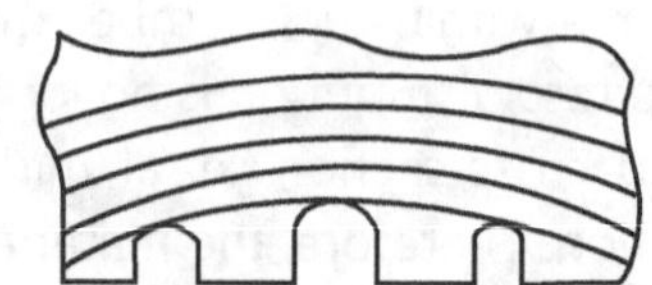

Good : Addition of two more notches

Fig. 2.23

2.

Poor

Good : Provision of narrow
projection instead of wide one

Fig. 2.24

3.

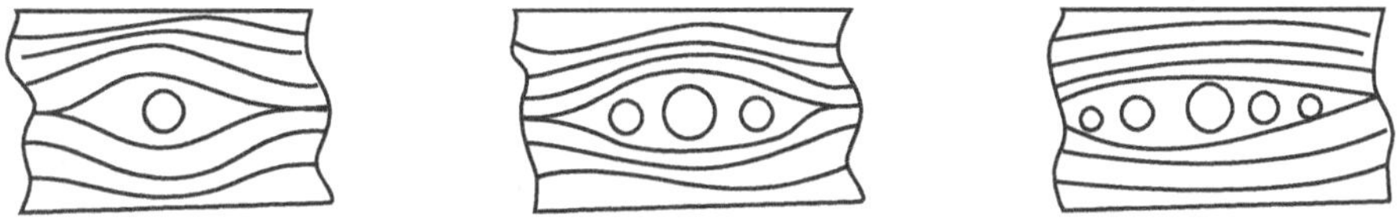

Poor   Good   Preferred

Addition of some more holes

Fig. 2.25

4.

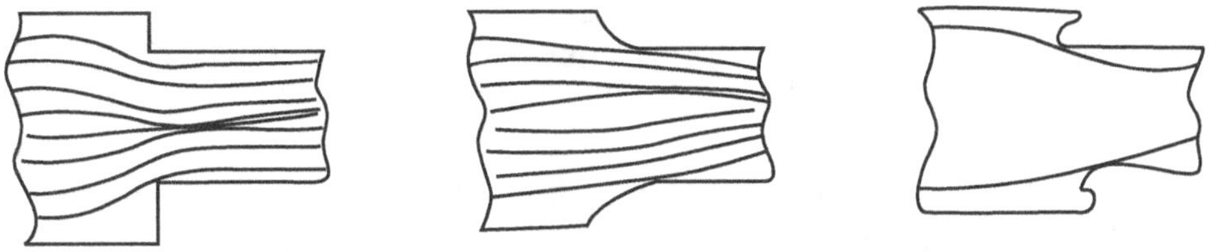

Poor : Sharp corner   Good : Fillet   Preferred : Undercut fillet

Fig. 2.26

5.

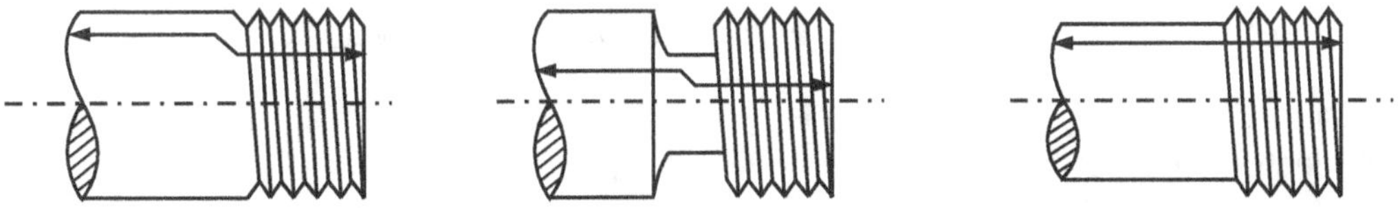

Poor   Good : Undercut   Preferred : Reduction in
shank diameter

Fig. 2.27

2.7 BOLTS OF UNIFORM STRENGTH   (S-16, 18; W-18)

Q.1. *What is the bolts of uniform strength ?*

In certain applications, bolts such as connecting rod bolts, bolts associated with presses, power hammers etc. are subjected to short or impact loads. In these, resilience of the bolt is important design consideration to prevent breakage at the threads. Resilience is the spring property of the bolt.

In case of ordinary bolt, the larger part of the impact energy will be absorbed in the threaded portion, since the energy absorbed per unit volume is proportional to the square of the stress, and the stress induced is higher in the threaded portion.

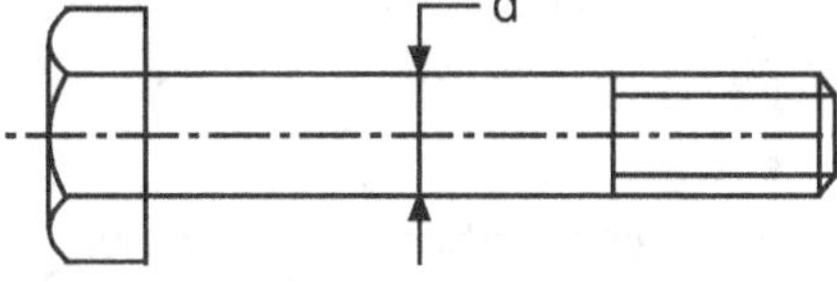

Fig. 2.28

Methods to Make a Bolt of Uniform Strength :

Following are the methods to make a bolt of uniform strength :

(a)　If the shank diameter of the bolt is made equal to that of the root diameter, then the stresses induced at all the cross-section of the bolt will be equal and so the energy absorbed is uniform throughout the length of the bolt. It is then called as a bolt of uniform strength.

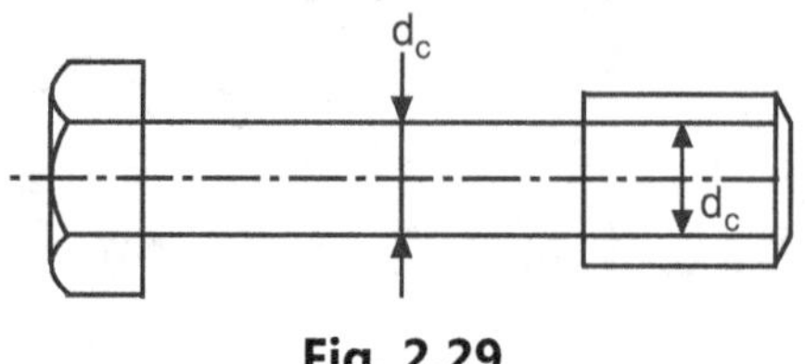

Fig. 2.29

where,　　　　　　　　d_c = core diameter

(b)　Instead of reducing the shank diameter, the shank area of the bolt can be made equal to that of the root area of the threads, by drilling a hole, upto threaded end as shown in Fig. 2.30.

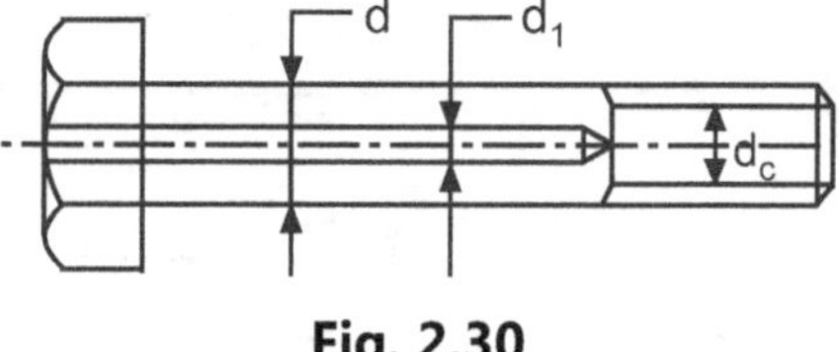

Fig. 2.30

Let,　　　　　　d_1 = diameter of the hole to be drilled

　　　　　　d = nominal diameter of bolt

　　　　　　d_c = core diameter of bolt.

Then, for a bolt of uniform strength,

$$\boxed{d_1 = \sqrt{d^2 - d_c^2}}$$

2.8 LOAD FACTOR, SERVICE FACTOR AND THEIR APPLICATIONS　　(W-16)

1.　Service Factor (SF)

Service factor is described as 'the service limit of the component for definite period of cycle'. Overload capacity is considered while designing component, device, engine, motor etc., as a safety factor. It is expressed usually by a number greater than one. For example, a SF of 1.15 means the item can take 15 % more load than its rated capacity without breakdown.

This means that a 10 hp motor with a 1.15 SF could provide 11.5 hp when required for short-term use. Some fractional horsepower motors have higher service factors, such as 1.25, 1.35 and even 1.50. In general, it is not a good practice to size motors to operate continuously above rated load in the service factor area.

2.　Velocity Factor :

The velocity factor accounts for :

- Factors such as pitch line velocity, manufacturing and assembly accuracies.
- The severity of impact as successive pairs of teeth comes into engagement.
- Polar mass moments of inertia of pinion and gear mesh shaft and bearing stiffness.
- The velocity factor is used to take care of possibilities of fatigue failure.

3.　Overload Factor :

The overload factor makes allowance for the externally applied loads which are in excess of the nominal tangential load. In determining the overload factor, consideration should be given to the fact that many prime movers and driven equipment, individually or in combination, develop momentary peak torques appreciably greater than those determined by the nominal ratings of either the prime mover or the driven equipment.

There are many possible sources of overload which should be considered. Some of these are: system vibrations, acceleration torques, over speeds, variations in system operation, split path load sharing among multiple prime movers, and changes in process load conditions.

Practice Questions

1. Name the type of failure in the following machine elements :
   - (i) Transmission shafts
   - (ii) Gears
   - (iii) Ball bearings
   - (iv) Springs
   - (v) Clutches.
2. State different types of loads.
3. Define stress. Explain types of stresses.
4. Draw stress-strain diagram for following material and mention all the points on it.
   (i) Ductile material, (ii) Brittle material.
5. Define factor of safety.
6. Explain maximum principle stress theory or Rankine's theory.
7. Explain maximum shear stress theory or Guests or Tresca's theory.
8. Define "stress concentration". Enlist methods to reduce stress concentration.
9. Define bolt of uniform strength.
10. Define load factor, service factor with their applications.

MSBTE Questions and Answers (As per G-Scheme)

Summer 2016

1. Describe the concept of "Bolts of Uniform Strength". **(4 M)**

Ans. Refer Article 2.6.1.

2. State any four criterias for selection of "Factor of Safety". **(4 M)**

Ans. Refer Article 2.3.2.

3. Describe service factor, overload factor, velocity factor and factor of safety. **(8 M)**

Ans. Refer Article 2.7.

Winter 2016

1. Define the terms : (i) Fatigue and (ii) Endurance limit with suitable example. **(4 M)**

Ans. Refer Article 2.5.

2. List No. of different factors to be considered to find design load and explain any two of them. **(4 M)**

Ans. Refer Article 2.8.

3. Explain maximum principal stress theory of failure. **(4 M)**

Ans. Refer Article 2.4 (1).

4. Draw and explain the stress-strain diagram for ductile material. **(4 M)**

Ans. Refer Article 2.2 (A).

Summer 2017

1. Define the following terms : (i) Factor of safety, (ii) Endurance limit. **(4 M)**

Ans. Refer Articles 2.3 and 2.5.

2. State the importance of S-N diagram for variable stresses. **(4 M)**

Ans. Refer Article 2.5.1.

3. Enlist any four factors affecting selection of factor of safety. **(4 M)**

Ans. Refer Article 2.3.2.

4. Explain maximum principle stress theory. **(4 M)**

Ans. Refer Article 2.4 (1).

Winter 2017

1. Define factor of safety. State the factors affecting its selection. **(4 M)**

Ans. Refer Articles 2.3.1 and 2.3.2.

2. Define the term : (i) Fatigue and (ii) Endurance limit. **(4 M)**

Ans. Refer Article 2.5.

3. Explain maximum shear stress theory of failure. **(4 M)**

Ans. Refer Article 2.4 (2).

4. Define stress concentration. State its causes. Explain the different methods to reduce stress concentration with suitable example. **(6 M)**

Ans. Refer Articles 2.6 and 2.6.3.

Summer 2018

1. Describe the concept of "Bolts of Uniform Strength". **(4 M)**

Ans. Refer Article 2.7.

2. Draw stress strain diagram for ductile material and state its importance. **(4 M)**

Ans. Refer Article 2.2 (A).

3. Explain - Maximum principal stress theory. **(4 M)**

Ans. Refer Article 2.4 (1).

4. Define factor of safety. State the factors affecting its selection. **(4 M)**

Ans. Refer Articles 2.3.1 and 2.3.2.

Winter 2018

1. List the important factors that influence the magnitude of F.O.S. **(4 M)**

Ans. Refer Article 2.3.2.

2. Define fatigue and endurance limit. Draw the S-N curve for cyclic loading. **(4 M)**

Ans. Refer Articles 2.5 and 2.5.1.

3. Explain the two methods to make bolt of uniform strength. **(4 M)**

Ans. Refer Article 2.7.

4. Draw stress-strain diagram for ductile material and state its importance. **(4 M)**

Ans. Refer Article 2.2.

✍ ✍ ✍

Chapter 3

DESIGN OF CHASSIS COMPONENTS

Weightage of Marks = 16, Teaching Hours = 16

Syllabus

3.1   Function of Tie Rod, Materials for Tie Rod with Justification and Design of Tie Rod

3.2   Function of Clutch, Material for Friction Lining with Justification, Design of Disc Clutch and Multi-plate Clutch Considering Uniform Wear Condition

3.3   Function of Propeller Shaft, Design of Propeller Shaft Including Universal Coupling

3.4   Function of Semi-elliptical Leaf Spring, Materials of Leaf Spring with Justifications and Design of Semi-elliptical Leaf Spring

About this Chapter

After reading this chapter, students will be able to :

- Choose suitable materials for the given chassis of component with justification.
- Explain stepwise design procedure for the given chassis component.
- Calculate dimensions of the given chassis component from the given data.
- Draw proportionate diagram of the given chassis component.

3.1 TIE ROD

Function of Tie Rod :

Tie rods are attached on both ends of the steering rack and as the pinion rolls over the slotted rack, they help to push and pull the front tyres as the steering wheel is turned.

Tie rods offer an important function to a vehicle's steering and therefore a car's overall safety.

Materials for Tie Rod :

A typical tie rod end is made of a steel material, such as a carbon steel and a low alloy steel because of an impact in the event of a collision.

3.1.1 Tie Rod Design Procedure

(1) Natural Frequency :

A member (tie rod) vibrates while operating under a time varying continuous loads, but the vibrations subsides after some time and this is known as natural frequency. It is important for a components working frequency to operate below its natural frequency in order not to be excited into resonance.

Frequency of Tie rod; $f \text{ (Hz)} = \dfrac{1}{2\pi} \sqrt{\dfrac{K}{m}}$

$$\quad \text{... (3.1)}$$

Where K is the stiffness constant and is given as

$$K = \frac{AE}{l} \quad \text{... (3.2)}$$

where,

f is the frequency in Hertz

m is the mass (load) in kg

E is the modulus of elasticity

l is length of tie rod, and A is the cross sectional area of the tie rod.

(2) Fracture Toughness :

Fracture Toughness (K_1C) can be obtained by following equation,

$$K_1C = \frac{wl}{A} (\pi c)^{0.5}$$

$$\text{... (3.3)}$$

Where, A is given as,

$$A = \frac{wl}{K_1C} (\pi c)^{0.5}$$

$$\text{... (3.4)}$$

Where,

K_1C is the fracture toughness

σ is the stress induced on the tie rod

C is a very small crack

wl is the total load on the tie rod.

(3) Fatigue Strength :

Considering the fatigue strength of the tie rod, it is important that the Fatigue strength endurance limit (σ_e) be as high as possible to enhance longevity of the tie rod during operation.

Following equation can be considered :

$$\left. \begin{array}{c} \dfrac{wl}{A} \leq \sigma_e \\[2mm] A \geq \dfrac{wl}{\sigma_e} \end{array} \right\}$$

$$\text{... (3.5)}$$

(4) Buckling Load :

If a slender bar such as tie rod of constant cross-section is pined at each end, the applied compressive load P_{cr}, that will result in buckling is expressed as,

$$P_{cr} = \frac{\pi^2 EI}{l^2}$$

$$\text{... (3.6)}$$

Where, I is the moment of inertia of cross-sectional area about an axis through the centroid.

l is the length of the tie rod.

$$\therefore \qquad P_{cr} = \frac{C\pi^2 EI}{l^2}$$

$$\text{... (3.7)}$$

In this case, the end condition constant (C) depends on the end conditions of a column which can be pinned-pinned, fixed-free, fixed-pinned and fixed-fixed, of which a tie rod end may fall into one of this categories.

Using the relation, $\qquad I = AK^2$

$\therefore$ Equation (3.7) can be written as,

$$\frac{P_{cr}}{A} = \frac{C\pi^2 E}{\left(\dfrac{l}{K}\right)^2}$$

$$\text{... (3.8)}$$

Where, $\dfrac{l}{K}$ is slenderness ratio

K is radius of gyration

$\dfrac{P_{cr}}{A}$ is the critical unit load.

Equation (3.8) shows that the critical unit load is mainly depend on the modulus of elasticity and the slenderness ratio. Hence, a column obeying the Euler formula manufactured with high strength alloy steel is not stronger than the one manufactured with low carbon steel, provided, E is same for both.

(5) Bending :

Bending is one of the failure modes associated to a straight material subjected to varying load conditions, of which, a vehicle tie rod is one of such materials due to its in-service condition.

Bending can be calculated from the bending equation :

$$\boxed{\dfrac{M}{I} = \dfrac{\sigma}{y}} \qquad \qquad \ldots (3.9)$$

Where, σ is the stress at a distance y from the neutral axis of the tie rod

M is the bending moment of the tie rod

y is the distance from the neutral axis

I is the second moment of area

In case, where the cross-section of the tie rod design is rectangular, the second moment of area can be expressed as,

$$\boxed{I = \dfrac{bh^3}{12}} \qquad \qquad \ldots (3.10)$$

Tie rod cross-section is cylindrical,

$$\therefore \qquad \qquad I = \dfrac{\pi d^4}{64} \qquad \qquad \ldots (3.11)$$

Where, d is diameter of the tie rod

h is the height of the tie rod

b is the width of the tie rod

3.2 DESIGN OF CLUTCH　　　　(S 15)

The use of clutch is found in Automobiles. It is used to connect a driving shaft to a driven shaft so that the driven shaft may be started or stopped without stopping the driving shaft.

In automobiles, friction clutch is used to connect the engine to the drive shaft.

Considerations in Designing a Friction Clutch :

Following considerations must be kept in mind while designing a friction clutch :

1. The moving parts of the clutch should have low weight.
2. The clutch should not require any external force to maintain contact of the friction surfaces.
3. Wear must be minimum between contact surfaces.
4. There should be heat dissipation from the contact surface.
5. The projecting parts must be protected by guard.

Materials for Friction Clutches :

Following materials are used for friction clutches :

1. Cast iron
2. Hardened steel.
3. Bronze on cast iron.
4. Powder metal on cast iron or steel.

Types of Friction Clutches :
1. Disc or plate clutches :
   (a) Single plate clutch,   (b) Multiplate clutch.
2. Cone clutches,
3. Centrifugal clutches.

3.2.1 Single Plate Clutch   (W-15)

> **Q.1.** *Derive the relation for torque to be transmitted by single plate clutch, considering uniform wear condition.*

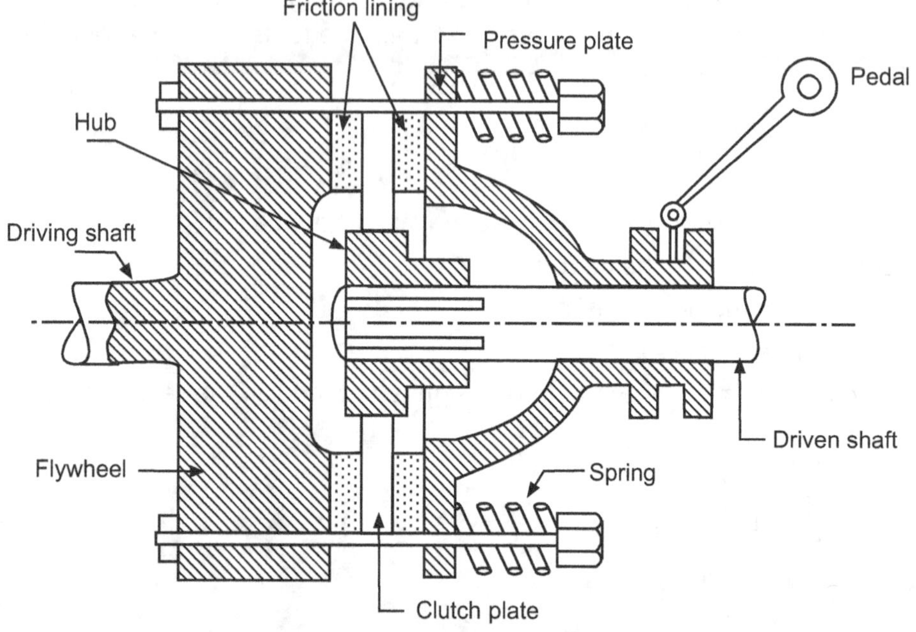

Fig. 3.1 : Single plate clutch

Design Procedure of Single Plate Clutch :   (S-15)

Consider two friction surfaces maintained in contact by an axial thrust (W) as shown in Fig. 3.2.

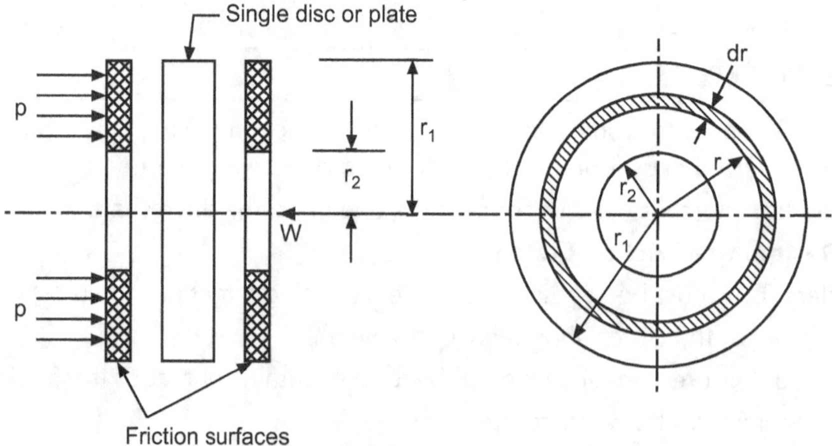

Fig. 3.2 : Details of axial thrust

Let,   T = Torque transmitted by the clutch

                    p = Intensity of axial pressure with which the contact surfaces are held together

r_1 and r_2 = External and internal radii of friction faces

                    r = Mean radius of the friction force

and μ = Coefficient of friction

Consider an elementary ring of radius r and thickness dr as shown in Fig. 3.2.

We know that, area of the contact surface or friction surface = $2\pi \cdot r \cdot dr$

Therefore, normal or axial force on the ring,

$$\delta w = \text{Pressure} \times \text{Area}$$
$$= p \times 2\pi \cdot r \cdot dr$$

and the frictional force on the ring acting tangentially at radius 'r' is,

$$F_r = \mu \times \delta w = \mu \cdot p \times 2\pi \cdot r \cdot dr$$

∴　　The frictional torque acting on the ring,

$$T_r = F_r \times r$$

∴　　　　　　$$T_r = \mu_p \times 2\pi r \cdot dr \times r$$

∴　　　　　　$$T_r = 2\pi\mu \cdot p \cdot r^2 \cdot dr$$

We shall now considering the following case :

(a)　When there is a uniform wear.

(b)　When there is a uniform pressure.

(a) When there is a uniform wear : (S-16, 17; W-16)

Let p be the normal intensity of pressure at a distance r from the axis of the clutch. Since, the intensity of pressure varies inversely with the distance, therefore,

$$p \cdot r = C$$　　　　　　　　　　...... where C = constant

∴　　　　　　$$p = \frac{C}{r}$$

And the normal force on the ring,

$$\delta w = p \cdot 2\pi r \cdot dr = \frac{C}{r} \times 2\pi r \cdot dr = 2\pi C \cdot dr$$

∴　　Total force acting on the friction surface,

$$W = \int_{r_2}^{r_1} 2\pi C dr = 2\pi C [r]_{r_2}^{r_1} = 2\pi C [r_1 - r_2]$$

∴　　　　　　$$C = \frac{W}{2\pi(r_1 - r_2)}$$

We know that, the frictional torque acting on the ring,

$$T_r = 2\pi\mu \cdot p \cdot r^2 \cdot dr = 2\pi\mu \times \frac{C}{r} \times r^2 \cdot dr = 2\pi\mu \cdot C \cdot r \cdot dr \qquad \dots \left(\because p = \frac{C}{r} \right)$$

∴　　Total frictional torque acting on the friction surface (or on the clutch),

$$T = \int_{r_2}^{r_1} 2\pi\mu \cdot C \cdot r \cdot dr$$

∴　　　　　　$$T = 2\pi\mu \cdot C \left[\frac{r^2}{2} \right]_{r_2}^{r_1}$$

∴　　　　　　$$T = 2\pi\mu \cdot C \left[\frac{(r_1)^2 - (r_2)^2}{2} \right]$$

∴　　　　　　$$T = \pi\mu \cdot C \left[(r_1)^2 - (r_2)^2 \right]$$

$$\therefore \qquad T = \pi\mu \times \frac{W}{2\pi(r_1 - r_2)}[(r_1)^2 - (r_2)^2]$$

$$\therefore \qquad T = \frac{1}{2} \times \mu \cdot W(r_1 + r_2)$$

$$\therefore \qquad \boxed{T = \mu \cdot W \cdot R}$$

where, $\qquad R = \dfrac{r_1 + r_2}{2}$ = Mean radius of the friction surface.

(b) When there is a uniform pressure condition :

When the pressure is uniformly distributed over the entire area of the friction face as shown in above figure, then intensity of pressure,

$$p = \frac{W}{\pi[(r_1)^2 - (r_2)^2]}$$

where, $\qquad W$ = Axial thrust with which the friction surfaces are held together

The frictional torque on elementary ring,

$$T_r = 2\pi\mu \cdot p \cdot r^2 \cdot dr$$

$\therefore \quad$ Total frictional torque acting on the friction surface or on the clutch,

$$T = \int_{r_2}^{r_1} 2\pi \cdot \mu \cdot p \cdot r^2 \cdot dr = 2\pi \cdot \mu \cdot p \cdot \left[\frac{r^3}{3}\right]_{r_2}^{r_1} = 2\pi \cdot \mu \cdot p \cdot \left[\frac{(r_1)^3 - (r_2)^3}{3}\right]$$

$$= 2\pi\mu \times \frac{W}{\pi[(r_1)^2 - (r_2)^2]}\left[\frac{(r_1)^3 - (r_2)^3}{3}\right] \qquad \dots \text{(Substituting the value of 'p')}$$

$$= \frac{2}{3} \cdot \mu \cdot W\left[\frac{(r_1)^3 - (r_2)^3}{(r_1)^2 - (r_2)^2}\right] = \mu \cdot W \cdot R$$

where, $\qquad R = \dfrac{2}{3}\left[\dfrac{(r_1)^3 - (r_2)^3}{(r_1)^2 - (r_2)^2}\right]$ = Mean radius of the friction surface.

SOLVED PROBLEMS

Problems on Single Plate Clutch :

Problem 3.1 : *A single plate clutch with both sides effective, has outer and inner diameter 300 mm and 200 mm respectively. The maximum intensity of pressure at any point in the contact surface not exceed 0.1 N/mm². If the coefficient of friction is 0.3, determine the power transmitted by a clutch at a speed of 2500 r.p.m.* **(W-16)**

Solution : Given :

Outer diameter, $\qquad d_1 = 300$ mm $\qquad\qquad \therefore r_1 = 150$ mm

Inner diameter, $\qquad d_2 = 200$ mm $\qquad\qquad \therefore r_2 = 100$ mm

Maximum intensity of pressure, $p_{max} = 0.1$ N/mm²

$$\mu = 0.3$$

$$P = \text{Power transmitted} = ?$$

$$N = 2500 \text{ rpm}$$

Considering uniform wear condition,

The intensity of pressure is at r_2

$$C = 0.1\, r_2 = 0.1 \times 100 = 10$$

$$W = 2\pi C(r_1 - r_2) = 2\pi \times 10(150 - 100) = \mathbf{3140\ N}$$

Torque transmitted, $T = n \times \mu \times W \times R$

where, $R = \dfrac{r_1 + r_2}{2} = \dfrac{150 + 100}{2} = 175$ mm

$\therefore$ $T = 2 \times 0.3 \times 3140 \times 175 = 329.7 \times 10^3$ Watt $= 329.7$ kW

Power transmitted, $P = \dfrac{2\pi \times N \times T}{60} = \dfrac{2\pi \times 2500 \times 329.7 \times 10^3}{60} = 862.72 \times 10^5$ Watt

$\therefore$ Power transmitted by clutch,

$$P = \mathbf{86.27 \times 10^3 \ kW}$$

Problem 3.2 : *A single plate dry clutch transmits 7.5 kW at 900 rpm, the axial pressure is limited to 0.7 N/mm². If the coefficient of friction is 0.25, find :*

(i)   Mean radius and face width of friction lining assuming ratio of mean radius to face width as 4.

(ii)   Outer and inner diameter of the clutch plate.   **(W-17)**

Solution : Given : $n = 2$ (number of friction surfaces)

$P = 7.5$ kW $= 7.5 \times 10^3$ Watt

$N = 900$ rpm

Coefficient of friction, $\mu = 0.25$

Maximum intensity of pressure,

$p_{max} = 0.7$ N/mm²

Let r_1 and $r_2 =$ Outer and inner radius of frictional surfaces respectively,

$R =$ Mean radius of the friction lining in mm,

$b =$ Face width of friction lining

Ratio of mean radius to face width $= \dfrac{R}{b} = 4$

We know that, area of friction faces, $A = 2\pi R \cdot b$

$\therefore$ Normal or axial force acting on the friction surfaces,

$W = A \times p = 2\pi R \cdot b \times p$

Also, Torque transmitted,

$T = n \times \mu \times W \times R$ ... Considering uniform wear

$\therefore$ $T = n \times \mu \times (2\pi R \cdot b \times p) \times R = n \times \mu \times \left[2\pi R \times \left(\dfrac{R}{4}\right) \times p\right] \times R$

$\therefore$ $T = \left(\dfrac{\pi}{2}\right) \times n \times \mu \times p \times R^3 = \dfrac{\pi}{2} \times 2 \times 0.25 \times 0.7 \times R^3 = 0.5495 \ R^3$ ... (1)

Power transmitted, $P = \dfrac{2\pi NT}{60}$

$\therefore$ $7.5 \times 10^3 = \dfrac{2\pi \times 900 \times T}{60}$

$\therefore$ $450 \times 10^3 = 2\pi \times 900 \times T$

$\therefore$ $T = \dfrac{450 \times 10^3}{2\pi \times 900} = \dfrac{450 \times 10^3}{5652}$

$\therefore$ $T = 79.62$ N-m $= 79.62 \times 10^3$ N-mm   ... (2)

From equation (1) and (2), we get,

$79.62 \times 10^3 = 0.5495 \ R^3$

$\therefore$ $\dfrac{79.62 \times 10^3}{0.5495} = R^3$

$$\therefore \qquad R^3 = 144895.3594$$

$$\therefore \qquad R = 52.52 \cong 53 \text{ mm}$$

Now, Face width, $\qquad b = \dfrac{R}{4} = \dfrac{53}{4} = 13.25 = 14$ mm

Also, mean radius, $\qquad R = \dfrac{r_1 + r_2}{2}$

$$\therefore \qquad 53 \times 2 = r_1 + r_2$$

$$\therefore \qquad r_1 + r_2 = 106 \text{ mm} \qquad\qquad \dots (3)$$

and, $\qquad\qquad b = r_1 - r_2 = 14 \qquad\qquad \dots (4)$

Add equations (3) and (4),we get,

$$r_1 + r_2 = 106$$
$$\underline{+\ r_2 - r_2 = 14}$$
$$\therefore \qquad 2r_1 = 120$$

$$\therefore \qquad r_1 = \dfrac{120}{2}$$

$$\therefore \qquad r_1 = 60 \text{ mm} \qquad\qquad \dots (5)$$

Putting value of r_1 in equation (3), we get,

$$r_1 + r_2 = 106$$

$$\therefore \qquad 60 + r_2 = 106$$

$$\therefore \qquad r_2 = 106 - 60 = 46 \text{ mm}$$

$\therefore$ Outer diameter, $\quad d_1 = 2r_1 = 2 \times 60 = \textbf{120 mm}$

Inner diameter, $\quad d_2 = 2r_2 = 2 \times 46 = \textbf{92 mm}$

Mean value, $\qquad R = \textbf{53 mm}$

Face width, $\qquad b = \textbf{14 mm}$

Problem 3.3 : *A single plate dry clutch transmit 8 kW at 940 rpm, the axial pressure is limited to 0.7 N/mm². If coefficient of friction is 0.25, find :*

(a) Mean radius and face width of friction lining assuming ratio of mean radius to face width as 4 and

(b) Outer and inner radii of clutch plate.   **(W-18)**

Solution : Given Data : n = 2, Power P = 8 kW = 8000 W, N = 940 rpm.

Coefficient of friction, μ = 0.25

Maximum intensity of pressure, P_{max} = 0.7 N/mm^2

Let, r_2 and r_1 = Outer and inner radius of frictional surfaces respectively.

$\qquad\qquad r$ = Mean radius of the friction lining in mm

$\qquad\qquad b$ = Face width of friction lining

Ratio of mean radius to the face width, $\dfrac{r}{b}$ = 4.

We know that, Area of friction faces = $2\pi rb \times P$

Therefore, Normal or axial force acting on friction faces, W = A $\times$ P = $2\pi rb \times P$

Torque transmitted, $\qquad$ T = nμWr (uniform wear)

$$= n\mu(2\pi rb \times P)r = n\mu \left[2\pi r \times \left(\dfrac{r}{4}\right) \times P \right] r$$

$$= \left(\dfrac{\pi}{2}\right) n\mu P r^3 = \left(\dfrac{\pi}{2}\right) \times 2 \times 0.25 \times 0.7 \times r^3$$

$$= 0.5497\ r^3 \text{ N-mm} \qquad\qquad \dots (1)$$

Power transmitted, $P = \dfrac{2\pi NT}{60}$

$$8000 = \dfrac{2\pi \times 940 \times T}{60}$$

$\therefore$ $T = 81.27 \text{ N-m} = 81.27 \times 10^3 \text{ N-mm}$ ... (2)

From equations (1) and (2),

$\therefore$ $r^3 = \dfrac{81.27 \times 10^3}{0.5497}\, 147.845 \times 10^3$

$\therefore$ $r = 52.877 \text{ mm} = 53 \text{ mm approximately.}$

Now, face width of the friction lining, $b = \dfrac{r}{4} = \dfrac{53}{4} = 13.25$

We know that, $b = r_1 - r_2 = 13.25 \text{ mm}$ ... (3)

$$r = \dfrac{r_1 + r_2}{2}$$

$\therefore$ $r_1 + r_2 = 2r = 2 \times 53 = 106 \text{ mm}$ ... (4)

Equating equations (3) and (4),

$$r_1 = \textbf{59.625 mm} \quad \text{and} \quad r_2 = \textbf{46.375 mm}$$

Problem 3.4 : *Single plate dry clutch is to be designed to transmit 70,000 W at 3000 rpm, the external to internal radius of friction surface ratio is 1.25.*

Take μ = 0.3 and maximum axial pressure as 0.1 N/mm^2.

Solution : Given : $P = 70,000 \text{ W} = 70 \times 10^3 \text{ W}, \ N = 3000 \text{ rpm}$

$$\dfrac{r_1}{r_2} = 1.25, \ \mu = 0.3$$

Maximum axial pressure, $p_{max} = 0.1 \text{ N/mm}^2$

Let, $d_1 = $ Outer diameter of frictional surface

$d_2 = $ Inner diameter of frictional surface

r_1 and $r_2 = $ Corresponding radii (in mm) of frictional surface

Torque transmitted by clutch,

$$T = \dfrac{P \times 60}{2\pi N} = \dfrac{70,000 \times 60}{2\pi \times 3000} = \dfrac{4200 \times 10^3}{18840} = 223 \text{ N-m}$$

$\therefore$ $T = 223 \times 10^3 \text{ N-mm}$

For uniform wear, $p.r = C \text{ (constant)}$

Since, intensity of pressure is maximum at inner radius (r_2),

$$p_{max} \times r_2 = C$$

$\therefore$ $C = \textbf{0.1 r}_2 \textbf{ N/mm}$

Normal or axial load acting on the frictional surface,

$$W = 2\pi C\,(r_1 - r_2) = 2\pi \times 0.1\, r_2\,(1.25\, r_2 - r_2) = 0.157\,(r_2)^2$$

Also, mean radius, $R = \dfrac{r_1 + r_2}{2} = \dfrac{1.25\, r_2 + r_2}{2} = \dfrac{2.25\, r_2}{2}$

$\therefore$ Torque transmitted (T),

$$T = n \cdot \mu \times W \times R$$

$$\therefore \quad 223 \times 10^3 = 2 \times 0.3 \times 0.157 \, (r_2)^2 \times \frac{2.25 \, r_2}{2}$$

$$\therefore \quad 223 \times 10^3 = 0.6 \times 0.157 \, (r_2)^2 \times 1.125 \, r_2$$

$$\therefore \quad 223 \times 10^3 = 0.1059 \, r_2^3$$

$$\therefore \quad \frac{223 \times 10^3}{0.1059} = r_2^3$$

$$\therefore \quad r_2 = 128.174 \text{ mm} = 129 \text{ mm}$$

$$\therefore \quad r_1 = 1.25 \times r_2 = 1.25 \times 129 = \mathbf{161.25 \cong 162 \text{ mm}}$$

$\therefore$ Outer diameter, $d_1 = 2r_1 = 2 \times 162 = \mathbf{324 \text{ mm}}$

       Inner diameter, $d_2 = 2r_2 = 2 \times 129 = \mathbf{258 \text{ mm}}$

$$\therefore \quad C = 0.1 \, r_2 = 0.1 \times 129 = 12.9$$

$\therefore$ Axial thrust (W),

$$W = 2\pi C \, (r_1 - r_2)$$

$$\therefore \quad W = 2\pi \times 12.9 \, (324 - 258)$$

Axial force, $W = \mathbf{5346.7 \text{ N}}$

Problem 3.5 : *Determine the maximum and minimum pressure in a plate clutch when axial force is 5 kN. The inside and outside radii are 60 mm and 120 mm. Assume uniform wear condition.*

Solution : Given : $W = 5 \text{ kN} = 5000 \text{ N}, \; r_2 = 60 \text{ mm}, \; r_1 = 120 \text{ mm}$

Maximum Pressure :

Let, $p_{max} = $ Maximum pressure

Since, intensity of pressure is maximum at inner radius,

$$p_{max} \times r_2 = C \quad \text{or} \quad C = 60 \, p_{max}$$

$\therefore$ Total force on the contact surface (W),

$$5000 = 2\pi C \, (r_1 - r_2) = 2\pi \times 60 \, p_{max} \, (120 - 60)$$

$$\therefore \quad 5000 = 22608 \, p_{max}$$

$$\therefore \quad p_{max} = \frac{5000}{22608} = \mathbf{0.22 \text{ N/mm}^2}$$

Minimum Pressure :

Let, $p_{min} = $ Minimum pressure

$\therefore$ Intensity of pressure is minimum at outer radius, r_1

$$\therefore \quad p_{min} \times r_1 = C \quad \text{or} \quad C = 120 \, p_{min}$$

$\therefore$ Total force on the contact surface (W),

$$5000 = 2\pi C \, (r_1 - r_2) = 2\pi \times 120 \, p_{min} \, (120 - 60)$$

$$\therefore \quad 5000 = 45216 \, p_{min}$$

$$\therefore \quad \frac{5000}{45216} = p_{min}$$

$$\therefore \quad p_{min} = \mathbf{0.1105 \text{ N/mm}^2}$$

$$\therefore \quad p_{max} = \mathbf{0.22 \text{ N/mm}^2} \quad \text{and}$$

$$p_{min} = \mathbf{0.1105 \text{ N/mm}^2}$$

3.2.2 Design of Multiplate Clutch

Use :

Multiplate clutches are used when it is necessary to transmit large torque.

Example : Scooters, Bikes.

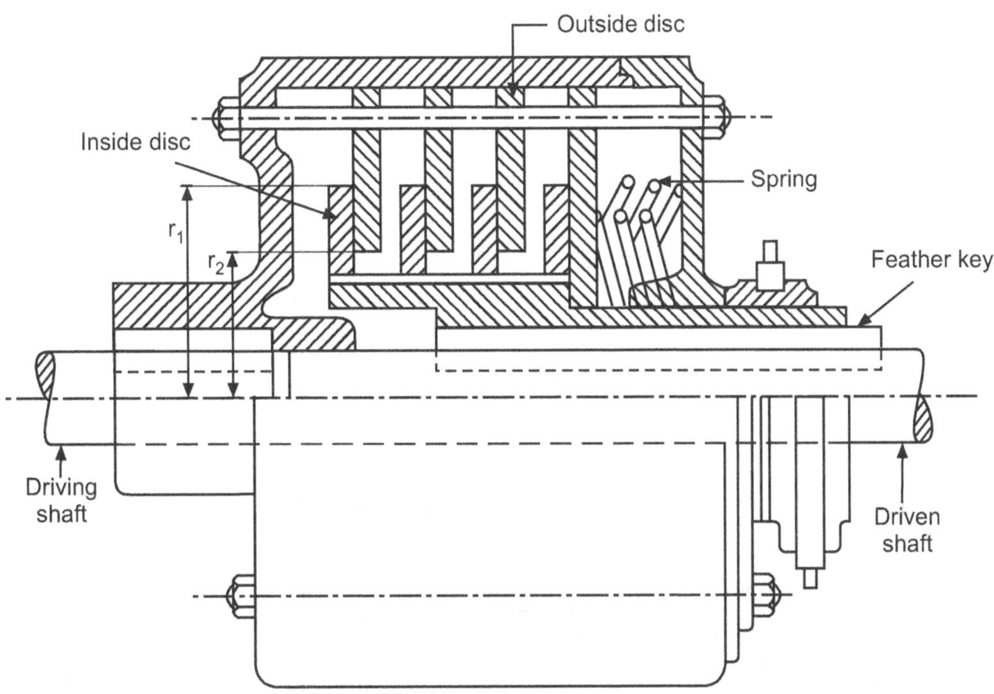

Fig. 3.3 : Multiplate clutch

Let, n_1 = Number of discs on the driving shaft

and n_2 = Number of discs on the driven shaft

$\therefore$ Number of pairs of contact surfaces,

$$n = n_1 + n_2 - 1$$

and total frictional torque acting on the friction surfaces or on the clutch,

$$T = n \cdot \mu \cdot W \cdot R$$

where,   R = Mean radius of friction surfaces

$$= \frac{2}{3}\left[\frac{(r_1)^3 - (r_2)^3}{(r_1)^2 - (r_2)^2}\right]$$ ... For uniform pressure

$$= \frac{r_1 + r_2}{2}$$ ... For uniform wear

Comparison of Single Plate Clutch and Multiplate Clutch :

Single Plate Clutch	Multiplate Clutch
1. In this, one clutch plate is used.	1. In this, more than two plates are used.
2. These type of clutches are used for higher axial forces and torque.	2. These type of clutches are used where large torque transmission is necessary.
3. E.g. Heavy vehicles, trucks, busses etc.	3. E.g. Scooters, motor cycles.

SOLVED PROBLEMS

Problems on Multiplate Clutch :

Problem 3.6 : *A multidisc clutch has 5 plates having 4 pairs of active friction surfaces, if the intensity of pressure is not be exceed 0.127 N/mm². Find the power transmitted at 500 rpm. The outer and the inner radii of friction surfaces are 125 mm and 75 mm respectively. Assume uniform wear and take coefficient of friction = 0.3.*

Solution : Given : $n_1 + n_2 = 5$, $n = 4$, $p = 0.127$ N/mm²,

$N = 500$ rpm, $r_1 = 125$ mm, $r_2 = 75$ mm, $\mu = 0.3$

For uniform wear, $\quad p.r = C$ (constant)

Since, intensity of pressure is maximum at inner radius (r_2)

$$p.r_2 = C$$

$$\therefore \quad C = 0.127 \times 75 = 9.525 \text{ N/mm}$$

$\therefore$ Axial thrust, $\quad W = 2\pi(r_1 - r_2) = 2\pi \times 9.525 (125 - 75) = 2993$ N

Mean radius of friction surfaces

$$R = \frac{r_1 + r_2}{2} = \frac{125 + 75}{2} = 100 \text{ mm} \cong \mathbf{0.1 \ m}$$

We know that,

Torque transmitted $\quad (T) = n \cdot \mu \cdot W \cdot R = 4 \times 0.3 \times 2993 \times 0.1 = 359$ N-m

$\therefore$ Power transmitted,

$$P = \frac{2\pi NT}{60} = \frac{2\pi \times 500 \times 359}{60} = 18800 \text{ Watt} = \mathbf{18.8 \ kW}$$

Problem 3.7 : *A multidisc clutch has three discs on driving shaft and two on the driven shaft. The outside diameter of the contact surface is 240 mm and inside diameter is 120 mm. Assuring uniform wear and $\mu = 0.3$. Find the maximum axial intensity of pressure between the discs for transmitting 25000 W at 1575 rpm.*

Solution : Given : $n_1 = 3$, $n_2 = 2$, $d_1 = 240$ mm, $d_2 = 120$ mm, $\mu = 0.3$

$$p_{max} = ?$$

Power, $\quad P = 25 \times 10^3$ Watt

$\quad N = 1575$ rpm

We know that, Power transmitted,

$$P = \frac{2\pi NT}{60}$$

$$\therefore \quad 25 \times 10^3 = \frac{2\pi \times 1575 \times T}{60}$$

$$\therefore \quad 1500 \times 10^3 = 2\pi \times 1575 \times T$$

$$\therefore \quad T = \frac{1500 \times 10^3}{2\pi \times 1575} = \frac{1500 \times 10^3}{9891} = 151.65 \text{ N-m} = 151653 \text{ N-mm}$$

For uniform wear, Maximum intensity of pressure; $p_{max} \times r_2 = C$

$$\therefore \quad C = 60 \times p_{max} \quad \dots (1)$$

Axial force on each friction surface;

$$W = 2\pi C(r_1 - r_2) = 2\pi \times 60 \times p_{max} (120 - 60) = 22{,}608 \times p_{max} \quad \dots (2)$$

Now, for uniform wear,

Mean radius, $\quad R = \dfrac{r_1 + r_2}{2} = \dfrac{120 + 60}{2} = \dfrac{180}{2} = 90$ mm

Number of pairs of contact surfaces,

$$n = n_1 + n_2 - 1 = 3 + 2 - 1 = 4$$

$\therefore$ Torque transmitted,

$$T = n \cdot \mu \cdot W \cdot R$$

$\therefore$ $151653 = 4 \times 0.3 \times (22,608\ p_{max}) \times 90$

$\therefore$ $151653 = 2441664\ p_{max}$

$\therefore$ $p_{max} = \dfrac{151653}{2441664} = 0.0621\ N/mm^2$

$\therefore$ Maximum axial intensity of pressure,

$$p_{max} = \mathbf{0.062\ N/mm^2}$$

Problem 3.8 : *A multiple clutch is to transmit 4.5 kW at 750 rpm. The inner and outer radii of contact surfaces are 40 and 70 mm respectively. The co-efficient of friction is 0.1. The average intensity of pressure is 0.35 N/mm². Find total number of clutch plates, actual axial force required, actual average pressure and actual maximum pressure.*   **(S-16)**

Solution : Given : Power transmitted by clutch, P = 4.5 kW

Speed of clutch,   N = 750 rpm

Inner radius, r_2 = 40 mm

Outer radius, r_1 = 70 mm

Coefficient of friction, μ = 0.1

Average intensity of pressure,

$$P_{avg} = 0.35\ N/mm^2$$

Power transmitted by the clutch,

$$P = \frac{2\pi NT}{60}$$

$\therefore$ $4.5 \times 10^3 = \dfrac{2\pi \times 750}{60} \times T$

$\therefore$ $T = 57.29\ Nm = 57.29 \times 10^3\ N\text{-}mm$

For uniform wear, mean radius of friction surface,

$$R = \frac{r_1 + R_2}{2} = \frac{70 + 40}{2} = 55\ mm$$

Axial forces, $W = P_{avg} \times \pi \times (r_1^2 - r_2^2) = 0.35 \times \pi \times (70^2 - 40^2) = 3628.539\ N$

1. Total number of clutch plates :

Torque transmitted by the clutch,

$$T = n\mu Wr$$

$\therefore$ $57.29 \times 10^3 = n \times 0.1 \times 3628.539 \times 55$

$\therefore$ $n = 2.83$

$\therefore$ $n = 3$

Since the number of pairs of contact surfaces must be even, therefore we use four pairs of contact surfaces.

$$n = 4$$

2. Actual axial force required :

$$W_1 = \text{Actual axial force required}$$

Torque, $T = n\mu Wr$

$\therefore$ $57.29 \times 10^3 = 4 \times 0.1 \times W_1 \times 55$

$\therefore$ $W_1 = \mathbf{2604.09\ N}$

3. Actual average pressure :

$$P_{avg} = \frac{W_1}{\pi(r_1^2 - r_2^2)} = \frac{2604.09}{\pi(70^2 - 40^2)} = \textbf{0.251 N/mm}^2$$

4. Actual maximum pressure :

$$P_{max} = \frac{C}{r_2}$$

Here,

$$C = \frac{W_1}{\pi(r_1 - r_2)} = \frac{2604.09}{\pi(70 - 40)} = 27.63 \text{ N/mm}$$

$$\therefore \qquad P_{max} = \frac{27.63}{40} = 0.690 \text{ N/mm}^2$$

3.3 PROPELLER SHAFT   (W-15; S-17)

Q.1. *Why propeller shafts are generally made hollow ?*   (S-15)
Q.2. *Describe the design procedure of a hollow propeller shaft.*

Propeller shafts of road vehicles are sufficiently long and operate at high speed. Consequently, whirling may occur at certain critical speed. This causes bending stresses in material that are higher than shearing stress caused by transmitted torque.

The tendency of a propeller shaft to whirl should be reduced. The critical speed of shaft increases with decrease in weight. Hence, propeller shafts are made hollow which increases the moment of inertia of section and keeps the weight minimum.

Materials used are Alloy steel such as nickel, nickel chromium, chrome vanadium steel.

Justification : Propeller shaft transmits power from gear box to differential. So it requires high torsional strength and rigidity as well as material that can take a lot of fatigue.

3.3.1 Design Procedure of Propeller Shaft

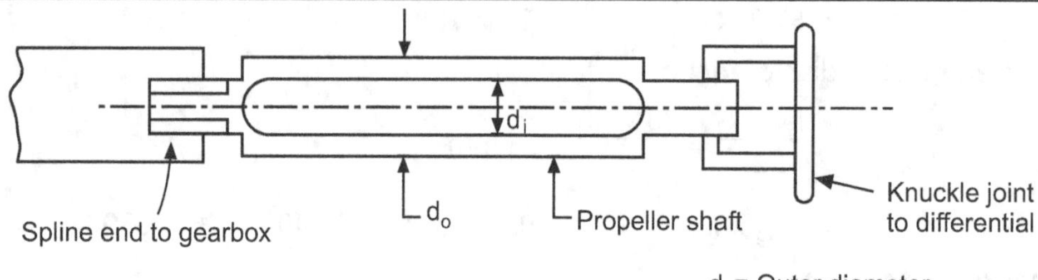

Fig. 3.4 : Propeller shaft

Find out dimensions, d_i = Inner diameter for shaft

                      d_o = Outer diameter for shaft

(a) Find out torque produced by the engine 'T_e'.

$$\therefore \qquad \boxed{P = \frac{2\pi N T_e}{60}}$$

where,   P = Power produced in kW or watt

           N = r.p.m.

          T_e = Torque produced by engine in N-m or N-mm.

(b) Torque transmitted by propeller shaft 'T_p'.

$$\therefore \qquad \boxed{T_p = T_e \times G_1}$$

where, G_1 = Gear reduction ratio

(c)   To find out d_i and d_o.

We know, $$K = \frac{d_i}{d_o}$$

Also we know that, $$T_p = \frac{\pi}{16} f_s (d_o)^3 (1 - K^4)$$

From above equation 'd_o' can be calculated.

3.3.2 Design of Couplings

Q.1.   *List different types of couplings and write requirement of good coupling.*

Shafts are usually available upto 7-8 metres in length. In order to have a greater length, shafts are connected to each other in two or more pieces of the shaft by means of a couplings.

Purpose of Couplings :

1.   To provide connection between shafts and alternator, generator etc.

2.   To provide connection between two misalignment shafts.

3.   To reduce transmission of shock loads from one shaft to another.

Requirements of Good Couplings :

1.   It should transmit full power from one shaft to the another.

2.   It should hold the shafts in perfect alignment.

3.   It should be easy to connect or disconnect.

4.   It should have no projection parts.

3.3.2.1 Universal Coupling (Hook's Coupling)

A universal or Hook's coupling is used to connect two shafts whose axes intersect at a small angle. The inclination of the two shafts may be constant, but in actual practice, it varies when the motion is transmitted from on shaft to another. The main application of the universal or Hook's coupling is found in the transmission from the gear box to the differential or back axle of the automobiles. In such a case, we use two Hook's couplings, one at each end of the propeller shaft, connecting the gear box at one end and the differential on the other end. A Hook's coupling is also used for transmission of power to different spindles of multiple drilling machines.

It is used as knee joint in milling machines. In designing a universal coupling, the shaft diameter and the pin diameter is obtained as discussed below. The other dimensions of the coupling are fixed by proportions as shown in Fig. 3.5.

Let d = Diameter of shaft, d

d_p = Diameter of pin,

τ and τ_1 = Allowable shear stress for the material of the shaft and pin respectively.

We known that torque transmitted by the shafts,

$$T = \frac{\pi}{16} \times \tau \times d^3$$

From the relation, the diameter of shafts may be determined. Since the pin is in double shear, therefore the torque transmitted,

$$T = 2 \times \frac{\pi}{4} (d_p)^2 \tau_1 \times d$$

From this relation, the diameter of pin may be determined.

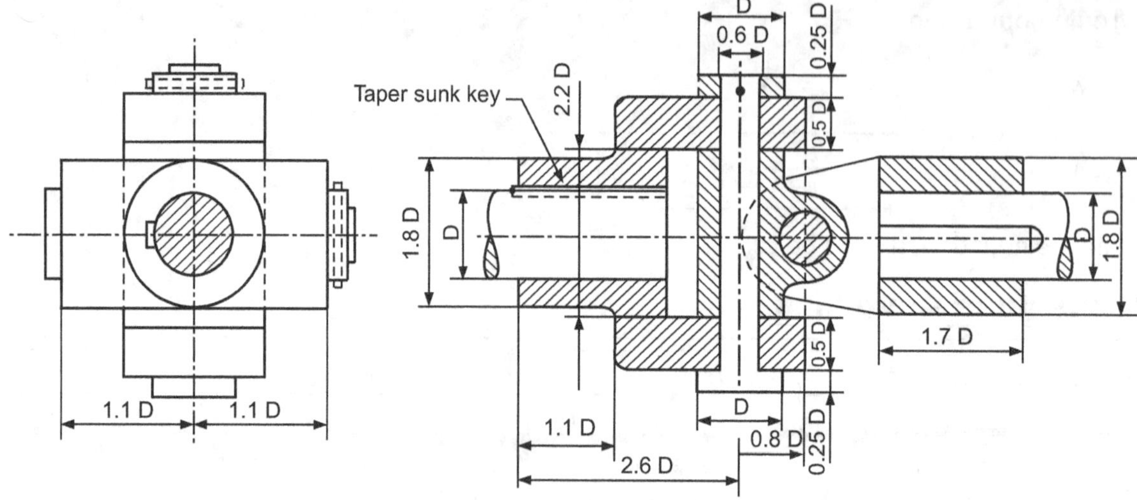

Fig. 3.5 : Universal coupling

SOLVED PROBLEMS

Problem 3.9 : *Design a hollow propeller shaft of a car with outside diameter 75 mm, transmits 22.5 kW at 1500 rpm to the wheels which are 90 cm in diameter. If the allowable shear stress is 60 N/mm². Find out (i) inner and outer diameter of shaft. (ii) cross-section of the shaft. Take gear box reduction ratio as 5.*

(W-14, 17; S-15)

Solution : Given :　　d_o = Outside diameter = 75 mm　　　N = 1500 rpm

P = 22.5 kW = 22.5×10^3 W　　　τ = Allowable shear stress = 60 N/mm²

G_1 = Gear reduction ratio = 5

Torque produced by engine (T_e)

We know,　　　　　　　$P = \dfrac{2\pi N T_e}{60}$

$\therefore$　　　　　$22.5 \times 10^3 = \dfrac{2 \times 3.14 \times 1500 \times T_e}{60}$

$\therefore$　　　　T_e = 143.24 N-m = $\mathbf{143.24 \times 10^3}$ **N-mm**

Now, torque transmitted by propeller shaft 'T_p'

$T_p = T_e \times G_1 = 143.24 \times 10^3 \times 5 = \mathbf{716.2 \times 10^3}$ **N-mm**

Let,　　　　　d_o = Outer diameter of propeller shaft

d_i = Inner diameter of propeller shaft

$K = \dfrac{d_i}{d_o} = \dfrac{d_i}{75}$

We know that,　　　$T_p = \dfrac{\pi}{16} f_s \, (d_o)^3 \, (1 - K^4)$

$\therefore$　　　$716.2 \times 10^3 = \dfrac{3.14}{16} \times 60 \times (75)^3 \, (1 - K^4)$

$\therefore$　　　　$1 - K^4 = 0.14$

$\therefore$　　　　$K^4 = 0.855$

　　　　$\dfrac{(d_i)^4}{(75)^4} = 0.855$

$\therefore$　　　　d_i = 72.1 mm $\cong$ 72 mm

$\therefore$　　　　d_o = **75 mm**　and　d_i = **72 mm**

Problem 3.10 : *Design a propeller shaft to transmit 5 kW at 5000 rpm with gear box reduction 16:1. Assume shear stress = 45 N/mm².*   **(S-18, W-16)**

Solution : Given : $P = 5 \text{ kW} = 5 \times 10^3$ Watt $G_1 = 16:1 =$ Gear reduction ratio

$N = 5000$ rpm $f_s = 45 \text{ N/mm}^2$

Torque produced by the engine (T_e)

We know, $P = \dfrac{2\pi N T_e}{60}$

$\therefore$ $5 \times 10^3 = \dfrac{2\pi \times 5000 \times T_e}{60}$

$\therefore$ $\dfrac{5 \times 10^3 \times 60}{2\pi \times 5000} = T_e$

$\therefore$ $\dfrac{300 \times 10^3}{31,400} = T_e$

$\therefore$ $T_e = \mathbf{9.554 \text{ N-mm}}$

Now, torque transmitted by propeller shaft (T_p),

$T_p = T_e \times G_1$ $(\ldots G_1 =$ gear reduction ratio)

$\therefore$ $T_p = 9.554 \times 16 = \mathbf{152.864 \text{ N-mm}}$

For solid shaft, $T_p = \dfrac{\pi}{16} \times f_s \times d^3$

$\therefore$ $152.864 = \dfrac{\pi}{16} \times 45 \times d^3$

$\therefore$ $d^3 = \dfrac{152.864 \times 16}{\pi \times 45} = \dfrac{2445.82}{141.3} = 17.309$

$\therefore$ Diameter of shaft, $d = \mathbf{2.58 \text{ mm}} \cong \mathbf{3 \text{ mm}}.$

Problem 3.11 : *Find the inside and outside diameter of a hollow shaft to transmit 20 kW at 200 rpm. The ultimate shear stress for the steel may be taken as 360 N/mm², factor of safety as 8 and the ratio of inside to outside diameters is 0.5..*

Solution : Given : $P = 20 \text{ kW} = 20 \times 10^3$ watt, $K = \dfrac{d_i}{d_o} = 0.5$, $N = 200$ rpm, $\tau_{ultimate} = 360 \text{ N/mm}^2$,

F.S. = 8. **Find :** d = Diameter of shaft

We know, allowable shear stress,

$\tau = \dfrac{\tau_{ultimate}}{F.S.} = \dfrac{360}{8} = 45 \text{ N/mm}^2$

Diameter of solid shaft   (d) = ?

Torque transmitted by the shaft,

$T = \dfrac{P \times 60}{2\pi N} = \dfrac{20 \times 10^3 \times 60}{2\pi \times 200} = \dfrac{1200 \times 10^3}{1256} = 955.41 \text{ N-m}$

$\therefore$ $T = \mathbf{955.41 \times 10^3 \text{ N-mm}}$ ... (1)

Let, $d_i =$ Inside diameter

and $d_o =$ Outside diameter

We know, the torque transmitted by the hollow shaft,

$$T = 955 \times 10^3 \qquad \text{[From Equation (1)]}$$

$$\therefore \quad 955 \times 10^3 = \frac{\pi}{16} \times \tau(d_o)^3 \, (1 - K^4)$$

$$\therefore \quad 955 \times 10^3 = \frac{\pi}{16} \times 45 \, (d_o)^3 \, [1 - (0.5)^4]$$

$$\therefore \quad 955 \times 10^3 = 8.3 \, (d_o)^3$$

$$\therefore \quad (d_o)^3 = \frac{955 \times 10^3}{8.3}$$

$$\therefore \quad (d_o)^3 = 115060$$

$$\therefore \quad d_o = 48.6 \text{ mm} \quad \text{Say } \mathbf{50 \text{ mm}}$$

$$\text{and} \quad d_i = 0.5 \, d_o = 0.5 \times 50 = \mathbf{25 \text{ mm}}$$

Inside diameter (d_i) = **25 mm**,   Outside diameter, d_o = **50 mm**

3.4 LEAF SPRING

Q.1. *Why nipping is provided in leaf spring ?* (S-15)

Q.2. *What material should be used for leaf spring ?*

3.4.1 Concept of Nipping (S-16, 17, 18; W-16, 17)

Explanation :

When the central bolt holding the leaves is tightened the full length leaf bend back, as shown by dotted line. And will have an initial stress in opposite direction. The graduated leaves will have an initial stress in the same direction as that of the normal load. When the load is applied, the full length leaf gets relieved first, consequently the full length leaf will be stressed less than graduated leaf. The initial leaf between leaves may be so adjusted that under maximum load conditions, all the leaves are equally stressed. So for this reason, nipping is provided in leaf spring.

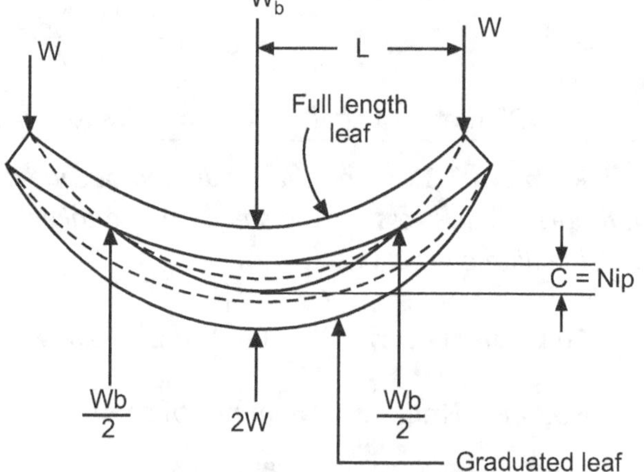

Fig. 3.6 : Process of nipping

Material :

According to Indian standards, for automobiles, plain carbon steel having 0.9 to 1.0% carbon, 50Cr1, 50Cr1V23, and 55Si2Mn90 all used in hardened and tempered state.

Reason to use plain carbon steel is, the leaves are heat treated after forming processes. The heat treatment of spring steel produces greater and therefore greater load capacity, greater range of deflection and better fatigue properties.

3.4.2 Design Procedure of Semi-elliptic Leaf Spring   (S-17, W-17)

Q.1. *Explain how a semi-elliptic leaf spring is designed ?*   (W-13, 14; S-15)

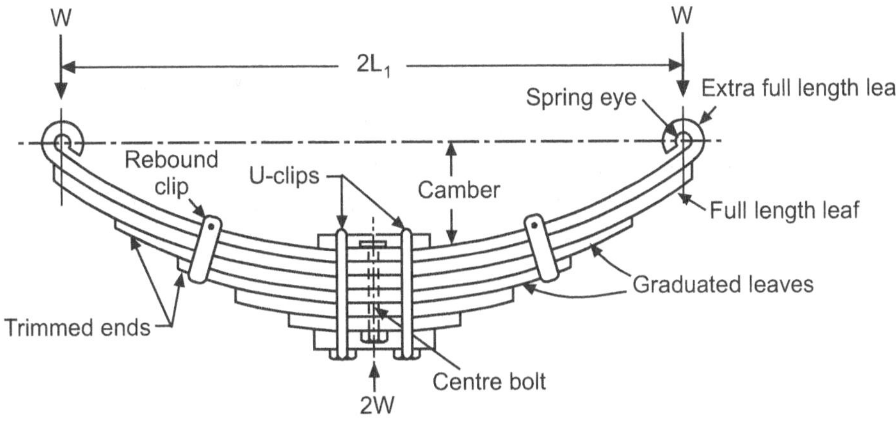

Fig. 3.7 : Semi-elliptical leaf spring   (S-16)

(a) Design steps for calculating thickness of leaf springs :

1. **Stress in leaf spring :**

$$\boxed{\sigma_b \;=\; \frac{\sigma WL}{nbt^2}}$$

where, effective length of spring $= 2L = 2L_1 - l$ (when central band is used)

$$= 2L = 2L_1 - \frac{2}{3}\cdot l \text{ (when 'U'-bolt is used)}$$

2. **Deflection in leaf spring :**

$$\delta \;=\; \frac{\sigma WL^3}{nEbt^3}$$

3. **Stress in full length leaves :**

$$\sigma_F \;=\; \frac{18WL}{bt^2\,(2n_g + 3n_f)}$$

4. **Stress in graduated leaves :**

$$\sigma_G \;=\; \frac{12WL}{bt^2\,(2n_g + 3n_f)}$$

5. **Deflection in full length and graduated leaves :**

$$\delta \;=\; \frac{12WL^3}{Ebt^2\,(2n_g + 3n_f)}$$

(b) Design steps for calculating length of leaf spring :

$$\text{Length of smallest leaf} \;=\; \frac{L \times 1}{(n-1)} + l$$

$$\text{Length of second smallest leaf} \;=\; \frac{(L \times 2)}{n-1} + l$$

$$\text{Length of } (n-1)^{\text{th}} \text{ leaf} \;=\; \frac{L \times (n-1)}{(n-1)} \times l$$

$$\text{Length of master leaf} \;=\; 2L_1 + \pi(d + t) \times 2$$

where,　　　　　　　　　　　　d = Diameter of eye

$$d = \left(\frac{32M}{\pi\sigma_b}\right)^{1/3}$$

An above equations,　　　　　　l = Ineffective length

　　　　　　　　　　　　　　L = Effective length

SOLVED PROBLEMS

Problem 3.12 : A semi-elliptical spring has an overall length of 1 m and sustain a load of 70 kN at its centre. The spring has 3 full length leaves and 15 graduated leaves with a central band of 100 mm width. All the leaves are to be stressed to 400 N/mm^2 when fully loaded. The ratio of total spring depth to that of the width is 2. Young modulus E = 0.2×10^6 N/mm^2. Determine :　　　　**(W-16)**

(1)　Thickness and width of the leaves.

(2)　Initial gap that should be provided between full length and graduated leaves before the band load is applied.

(3)　The load exerted on the band after the spring is assembled.

Solution : Given Data :　$2W$ = Central load = 70 kN,　　　L = 500 mm, l = 100 mm,

　　　　　　　　　$W = 35 \times 10^3$ N,　　　　　$n_f = 3,$　$n_g = 15,$

　　　Overall length of spring = 1 M,　　　　　$n = n_f + n_g = 18.$

Ratio of total spring depth to width = 2

$$\sigma_b = 400 \text{ N/mm}^2$$

Modulus of elasticity, $E = 2 \times 10^6$ N/mm^2

(1)　Thickness and width of leaves :

　　　　　　　　t = thickness of leaves, b = width of leaves

　　　　　　　$n = n_f + n_g = 15 + 3 = 18$

As ratio of spring depth ($n \times t$) to width of leaves is 2.

$$\therefore \quad \frac{n \times t}{b} = 2 \quad \therefore \quad \frac{18 \times t}{b} = 2 \quad \therefore \quad b = \mathbf{9t}$$

Effective length of leaves,　　　　$2L = 2L_1 - l = 1000 - 100 = 900$ mm

$$\therefore \quad L = \mathbf{450 \text{ mm}}$$

As all leaves are equally stressed,　　$\sigma_b = \dfrac{6\,WL}{nbt^2}$

$$\therefore \quad 400 = \frac{6 \times 35 \times 10^3 \times 450}{18 \times 9t \times t^2} = \frac{583 \times 10^3}{t^3}$$

$$t^3 = 1458 \quad \therefore \quad t = 11.34 \text{ mm} = 12 \text{ mm}$$

$$b = 9\,t = \mathbf{108 \text{ mm}}$$

(2)　Initial gap :　　　　　$C = \dfrac{2WL^3}{nEbt^3} = \dfrac{2 \times 35 \times 10^3 \times 450^3}{18 \times 0.2 \times 10^6 \times 108 \times 12^3}$

$$C = \mathbf{9.5 \text{ mm}}$$

(3)　The load exerted on the band after the spring is assembled.

$$W_b = \frac{2\,n_f \times n_g \times W}{n(2\,n_g + 3\,n_f)} = \frac{2 \times 3 \times 15 \times 35 \times 10^3}{18(2 \times 15 + 3 \times 3)}$$

$$W_b = \mathbf{4487 \text{ N}}$$

Problem 3.13 : *A truck spring has 12 numbers of leaves, two of which are full length leaves. The spring supports are 1.05 m apart and central band is 85 mm wide. The central load is to be 5.4 kN with a permissible stress of 280 N/mm². Determine the thickness and width of the steel spring leaves. The ratio of the total depth to the width of spring is 3. Also determine the deflection of the spring.* **(W-13, 18; S-15, 18)**

Solution : Given : $n = 12$, $n_F = 2$, $2L_1 = 1.05$ m $= 1050$ mm,

$l = 85$ mm, $2W = 5.4$ kN $= 5400$ N or $W = 2700$ N, $\sigma = 280$ N/mm²

1. Thickness (t) and width (b) of the spring leaves :

Since, it is given that, the ratio of the total depth of the spring (n × t) width of the spring (b) is 3,

$$\frac{n \times t}{b} = 3$$

$$\therefore \quad \frac{12 \times t}{b} = 3 \quad \text{or} \quad b = \frac{12t}{3} = 4t$$

We know that, the effective length of the spring,

$$2L = 2L_1 - l = 1050 - 85 = 965 \text{ mm}$$

$$L = \frac{965}{2} = 482.5 \text{ mm}$$

and the number of graduated leaves,

$$n_g = n - n_f = 12 - 2 = 10$$

Assuming that the leaves are not initially stressed, therefore maximum stress or bending stress for full length leaves,

$$\sigma = \frac{18 \, WL}{bt^2 \, (2n_g + 3n_f)}$$

$$\therefore \quad 280 = \frac{18 \times 2700 \times 482.5}{4t \times t^2 \, (2 \times 10 + 3 \times 2)} = \frac{225.476}{t^3}$$

$$\therefore \quad t^3 = \frac{225.476}{280} = 805.3$$

or $\qquad\qquad t = $ **9.3** say **10 mm**

and $\qquad\qquad b = 4t = 4 \times 10 = $ **40 mm**

Deflection of the spring :

$$\delta = \frac{12 \, W \cdot L^3}{Ebt^3 \, (2n_g + 3n_f)} = \frac{12 \times 2700 \times (482.5)^3}{0.21 \times 10^6 \times 40 \times (2 \times 10 + 3 \times 2)} \text{ mm}$$

$$\therefore \quad \delta = \textbf{16.7 mm} \text{ (Taking, } E = 0.21 \times 10^6 \text{ N/mm}^2)$$

Problem 3.14 : *A truck spring has 10 numbers of leaves. The supports are 1185 mm apart and the central (support) is 85 mm wide. The load on the spring is 20 kN and takes permissible stress of 300 N/mm². Determine the thickness of the leaves if the width of spring is 85 mm.* **(S-14; W-14)**

Solution : Given : $\qquad n = 10$, Assuming, number of full length leaves, $n_F = 2$

$$2L_1 = 1185 \text{ mm}$$

$$l = 85 \text{ mm}$$

$$2W = 20 \text{ kN} = 20000 \text{ N}$$

$$W = 10000 \text{ N}$$

$$\sigma = 300 \text{ N/mm}^2$$

Let, $\qquad\qquad t = $ Thickness of leaves

and, $\qquad\qquad b = $ Width of the leaves

We know that, the effective length of the spring

$$2L = 2L_1 - l = 1185 - 85 = 1100 \text{ mm}$$

$$\therefore \qquad L = \mathbf{550\ mm}$$

and, number of graduated leaves,

$$n_6 = n - n_f = 10 - 2 = 8$$

Assuming, the leaves are not initially stressed.

$\therefore$ The maximum bending stress for full length leaves (σ)

$$300 = \frac{18\ W \cdot L}{bt^2\ (2n_g + 3n_f)} = \frac{18 \times 10000 \times 550}{85 \times t^2\ (2 \times 8 + 3 \times 2)} = \frac{18 \times 10000 \times 550}{85 \times t^2\ (22)}$$

$$\therefore \qquad 300 = \frac{180000 \times 550}{1870\ t^2} = \frac{52941.176}{t^2}$$

$$\therefore \qquad t^2 = \frac{52941.176}{300} = 176.47$$

$$\therefore \qquad t = \mathbf{13.28\ mm} \quad \text{Say } t = \mathbf{14\ mm}$$

OR

$$f = \frac{6\ WL}{nbt^2}$$

$$300 = \frac{6 \times 10000 \times 550}{10 \times 85 \times t^2}$$

$$\therefore \qquad t = \mathbf{11.37\ mm} \quad \text{Say } t = \mathbf{12\ mm} \text{ (approx.)}$$

Problem 3.15 : *A truck spring has 12 number of leaves, two of which are full length leaves. The spring supports are 1.10 m apart and the central band is 90 mm wide. The central load is 6 kN with permissible stress of 300 MPa. Determine thickness and width of spring leaves. If ratio of total depth to width of the spring is 3. Also determine the deflection of the spring.*

Solution : Given : $n = 12$, $n_F = 2$, $2L_1 = 1.10$ m $= 1100$ mm, $l = 90$ mm,

$$2W = 6 \text{ kN} = 6000 \text{ N} \qquad \therefore\ W = 3000 \text{ N}$$

$$\sigma = 300 \text{ MPa} = 300 \text{ N/mm}^2.$$

1. **Thickness (t) and width (b) of the leaf spring :**

    The ratio of total depth ($n \times t$) and width of spring (b) is 3.

    $$\therefore \qquad \frac{n \times t}{b} = 3$$

    $$\therefore \qquad \frac{12 \times t}{b} = 3$$

    $$\therefore \qquad b = 4t$$

    We know, the effective length of the spring,

    $$2L = 2L_1 - l = 1100 - 90 = 1010$$

    $$\therefore \qquad L = \frac{1010}{2} = 505 \text{ mm}$$

    and, number of graduated leaves,

    $$n_g = n - n_f = 12 - 2 = 10$$

Assuming that, the leaves are not initially stressed, therefore maximum stress or bending stress for full length leaves (σ).

$$\therefore \qquad 300 = \frac{18\ WL}{bt^2\ (2n_g + 3n_f)} = \frac{18\ WL}{4t \times t^2\ (2 \times 10 + 3 \times 2)} = \frac{18 \times 3000 \times 505}{4t^3\ (26)}$$

$$\therefore \qquad t^3 = \frac{18 \times 3000 \times 505}{4 \times 26 \times 300} = \frac{18 \times 3000 \times 505}{31,200}$$

$$\therefore \qquad t^3 = 874.04$$

$$\therefore \qquad t = \mathbf{9.56\ mm} \quad \text{Say } t = \mathbf{10\ mm}$$

$$\therefore \qquad b = 4t = 4 \times 10 = \mathbf{40\ mm}$$

Deflection of the Spring :

$$\delta = \frac{12\,W \cdot L^3}{Ebt^3\,(2n_g + 3n_f)} = \frac{12 \times 3000 \times (505)^3}{0.21 \times 10^6 \times 40 \times (10)^3\,(2 \times 10 + 3 \times 2)}$$

$$\text{... (Assume } E = 0.21 \times 10^6 \text{ N/mm}^2)$$

$$\therefore \quad \delta = \frac{12 \times 3000 \times (505)^3}{2,18,400 \times 10^6} = \frac{4.636 \times 10^{12}}{2,18,400 \times 10^6} = 21.22 \text{ mm}$$

$\therefore$ Deflection of the spring, δ = **21.22 mm**

Problem 3.16 : *A semi-elliptical leaf spring consists of two full length leaves and eight graduated leaves including master leaf. The effective length of the spring is 1 m and maximum force acting on it is 10 kN, width of each leaf is 50 mm. The spring is initially preloaded so that stresses induced in each leaf are 350 N/mm². If modulus of elasticity of spring material is 207000 N/mm², determine thickness of each leaf, deflection of spring and initial nip. Sketch proportionate figure of semi-elliptical leaf spring.* **(W-15)**

Solution : Given : Maximum force = 2P = 10 kN; P = 5000 N;

Effective length of the spring, 2L = 1 M = 1000 mm;

$\quad$ L = 500 mm; $\quad n_f = 2$; $\quad n_g = 8$; $\quad n = n_f + n_g = 2 + 8 = 10$;

Width, $\qquad\qquad\qquad$ b = 50 mm;

Modulus of elasticity, $\qquad$ E = 207000 N/mm²;

$\qquad\qquad\qquad\qquad$ σ_b = 350 N/mm²

1. Thickness of leaves :

Since the stresses are equal in all leaves,

$$\sigma_b = \frac{6\,PL}{nbt^2}$$

$$\therefore \qquad 650 = \frac{6 \times 5000 \times 500}{10 \times 50 \times t^2}$$

$\therefore$ Thickness, $\qquad$ t = **9.26 mm** $\cong$ **10 mm**

2. Deflection of spring :

$$\delta = \frac{12\,PL^3}{Ebt^3\,(3n_f + 2n_g)} = \frac{12 \times 5000 \times 500^3}{207000 \times 50 \times 10^3\,(3 \times 2 + 2 \times 8)} = \textbf{32.93 mm}$$

3. Initial nip :

$$C = \frac{2\,PL^3}{Enbt^3} = \frac{2 \times 5000 \times 500^3}{207000 \times 10 \times 50 \times 10^3} = \textbf{12.07 mm}$$

For Fig. : Refer Fig. 3.6.

3.4.3 Comparison between Semi-Elliptical Leaf Spring and Helical Spring

Q.1. *Compare semi-elliptic leaf spring and helical spring used in automotive suspension system with respect to : (i) Materials, (ii) Strength, (iii) Comfort.*

Sr. No.	Points	Semielliptic Leaf Spring	Helical Spring
1.	Materials	Plain carbon steel, 50 Cr_1, $50Cr_1$ V_{23}, 55 Si_2 Mn_{90}	Oil tempered carbon steel containing 0.60 to 0.70% carbon and 0.60 to 1.0% manganese.
2.	Strength	Strength is more used on heavy duty.	Strength is less than leaf spring.
3.	Comfort	Less comfort obtained.	Comfort is more than semi-elliptical leaf spring.

3.4.4 Types of Leaves Used in Engineering Practice with their Applications

(S-15)

Types of Leaves :
1. Full length leave.
   (a) Full length leave with eye.   (b) Full length leave without eye.
2. Graduated leave.

Applications of Leaves :
1. It is used in semi-elliptical leaf spring.   2. It is used in quarter elliptical leaf spring.
3. It is used in three quarter elliptical leaf spring.   4. It is used in full elliptical leaf spring.

Problems for Practice

1. Compare semi-elliptical leaf spring and helical spring used in automotive suspension system with respect to – (i) Material, (ii) Strength, (iii) Comfort.   **(S-14)**
2. Explain design procedure of leaf spring.   **(W-13)**
3. Why nipping is provided in leaf spring.
4. Describe design procedure of hollow propeller shaft.

Problems for Practice

Problems on Clutch Plate :

1. A single plate clutch with both sides of the plate effective is required to transmit 25 kW at 1600 rpm. The outer diameter of the plate is limited to 300 mm and the intensity of pressure between the plates not to exceed 0.07 N/mm^2. Assuming uniform wear and coefficient of friction 0.3, find the inner diameter of the plates and the axial force necessary to engage the clutch.

   (Ans. d_i = 90 mm, p = 2374 N**)**

2. A multiple disc clutch has three discs on the driving shaft and two on the driven shaft, providing four pairs of contact surfaces. The outer diameter of the contact surfaces is 250 mm and the inner diameter is 150 mm. Determine the maximum axial intensity of pressure between the discs for transmitting 18.70 kW at 500 rpm. Assume uniform wear and coefficient of friction as 0.3.

Problems on Leaf Spring :

3. A locomotive semi-elliptical laminated spring has an overall length of 1 m and sustain a load of 70 kN at its centre. The spring has 3 full length leaves and 15 graduated leaves with a central band of 100 mm width. All the leaves are to be stressed to 400 MPa, when fully loaded.
   The ratio of the total spring depth to that of width is 2.
   $$E = 210 \text{ kN/mm}^2$$
   Determine : (1) The thickness and width of the leaves.   (2) Deflection of spring.

4. A semi-elliptical laminated spring 900 mm long and 55 mm wide is held together at the centre by a band 50 mm wide. If the thickness of each leaf is 5 mm, find the number of leaves required to carry a load of 4200 N. Assume a maximum working stress of 480 MPa.
   If the two of these leaves extend the full length of the spring, find the deflection of the spring. The Young's modulus for the spring material may be taken at 210 kN/mm^2.

   (Ans. No. of leaves = 9, δ = 72 mm**)**

MSBTE Questions and Answers (As per G-Scheme)

Summer 2016

1. State the design procedure of single plate clutch using wear condition.   **(4 M)**
   Ans. Refer Article 3.2.1 (a).
2. Explain concept of nipping.   **(4 M)**
   Ans. Refer Article 3.4.1.

3. Draw neat sketch of leaf spring and show span of spring, ineffective length, central load, width of spring, total depth of spring. **(4 M)**

Ans. Refer Article 3.4.2.

4. A multiple clutch is to transmit 4.5 kW at 750 rpm. The inner and outer radii of contact surfaces are 40 and 70 mm respectively. The co-efficient of friction is 0.1. The average intensity of pressure is 0.35 N/mm^2. Find total number of clutch plates, actual axial force required, actual average pressure and actual maximum pressure. **(8 M)**

Ans. Refer Problem 3.8.

<hr>

Winter 2016

1. Derive the relation for torque to be transmitted by single plate clutch considering uniform wear condition. **(4 M)**

Ans. Refer Article 3.2.1.

2. Design a propeller shaft to transmit 5 kW at 5000 r.p.m. with gear box reduction 16 : 1. Assume permissible shear stress for shaft material is 45 N/mm^2. **(4 M)**

Ans. Refer Solved Problem 3.10.

3. Describe the Nipping of leaf springs with neat sketch. **(4 M)**

Ans. Refer Article 3.4.

4. A single plate clutch with both side effective, has outer and inner diameter 300 mm and 200 mm respectively. The maximum intensity of pressure at any point in the contact surface is not exceed 0.1 N/mm^2. If the coefficient of friction is 0.3, determine the power transmitted by clutch at a speed of 2500 r.p.m. **(4 M)**

Ans. Refer Solved Problem 3.2.

5. A semi-elliptical spring has an overall length of 1 m, and sustain a load of 70 kN at its centre. The spring has 3 full length leaves and 15 graduated leaves with a central band of 100 mm width. All the leaves are to be stressed to 400 N/mm^2 when fully loaded. The ratio of total spring depth to that of the width is 2. Young modulus E = 0.2×10^6 N/mm^2. Determine :
   (i) Thickness and width of the leaves.
   (ii) Initial gap that should be provided between full length and graduated leaves before the band load is applied. **(8 M)**

Ans. Refer Solved Problem 3.12.

<hr>

Summer 2017

1. State the design procedure for semi-elliptical leaf spring. **(6 M)**

Ans. Refer Article 3.4.2.

2. Explain the concept of nipping. **(4 M)**

Ans. Refer Article 3.4.1.

3. State the design procedure for single plate clutch on the basis of uniform pressure theory. **(4 M)**

Ans. Refer Article 3.2.1.

4. State design procedure of propeller shaft. **(4 M)**

Ans. Refer Article 3.3.

<hr>

Winter 2017

1. A single plate dry clutch transmits 7.5 kW at 900 rpm. The axial pressure is 0.7 N/mm^2. Determine the outer and inner diameters of frictional surfaces if μ = 0.25. Take ratio of diameter as 1.25. Assume uniform wear theory. **(6 M)**

Ans. Refer Solved Problem 3.3.

2. Explain why nipping of leaf spring is necessary with neat sketch. **(4 M)**

Ans. Refer Article 3.4.1.

3. State the design considerations in semi-elliptical leaf spring. **(4 M)**

Ans. Refer Article 3.4.2.

4. Design a hollow propeller shaft of a car with outside diameter 75 mm, transmits 22.5 kW at 1500 rpm to the wheels which are 90 cm in diameter. If the allowable shear stress is 60 N/mm^2. Find out (i) inner and outer diameter of shaft. (ii) cross-section of the shaft. Take gear box reduction ratio as 5. **(8 M)**

Ans. Refer Solved Problem 3.9.

Summer 2018

1. Design a propeller shaft to transmit 5 kW at 5000 rpm, with gear box reduction 16 : 1. Assume permissible shear stress for shaft material is 45 N/mm^2. **(4 M)**

Ans. Refer Solved Problem 3.10.

2. Describe nipping of leaf springs with neat sketch. **(4 M)**

Ans. Refer Article 3.4.1.

3. A truck spring has 12 number of leaves, two of which are full length leaves. The spring supports are 1.05 m apart and central band is 85 mm wide. The central load is 5.4 kN with a permissible stress of 280 N/mm^2. Determine thickness and width of the steel spring leaves. The ratio of total depth to the width of the spring is 3. Also determine the deflection of the spring. **(8 M)**

Ans. Refer Solved Problem 3.13.

4. A multiple disc clutch has five plates having four pairs of active friction surfaces. If the intensity of pressure is not to exceed 0.127 N/mm^2. Find power transmitted at 500 rpm. The outer and inner radii of friction surfaces are 125 mm and 76 mm respectively. Assume uniform wear and take coefficient of friction = 0.3. **(8 M)**

Ans. Refer Similar Solved Problem 3.6.

Winter 2018

1. Design a propeller shaft to transmit 8 kW at 6500 rpm with gear box reduction 16 : 1. Assume shear stress f_s = 52 N/mm^2. **(4 M)**

Ans. Refer Similar Solved Problem 3.10.

2. A truck spring has 12 numbers of leaves, two of which are full length leaves. The spring supports are 1.05 m apart and central band is 85 mm wide. The central load is to be 5.4 kN with a permissible stress of 280 N/mm^2. Determine the thickness and width of steel spring leaves. The ratio of total depth to width of the spring is 3. Also determine the deflection of the spring. **(8 M)**

Ans. Refer Solved Problem 3.13.

3. A single plate dry clutch transmit 8 kW at 940 rpm, the axial pressure is limited to 0.7 N/mm^2. If coefficient of friction is 0.25, find :

　(a) Means radius and face width of friction lining assuming ratio of mean radius to face width as 4 and

　(b) Outer and inner radii of clutch plate. **(8 M)**

Ans. Refer Solved Problem 3.3.

4. A multi-disc clutch has 5 plates having 4 pairs of active friction surfaces, if the intensity of pressure is not to exceed 0.127 N/mm^2. Find power transmitted at 500 r.p.m. The outer and inner radii of friction surfaces are 130 mm and 80 mm respectively. Assume uniform wear and take coefficient of friction = 0.35. **(4 M)**

Ans. Refer Similar Solved Problem 3.6.

☝ ☝ ☝

Chapter 4

DESIGN OF ENGINE COMPONENTS

Weightage of Marks = 16, Teaching Hours = 18

Syllabus

4.1 Function of Cylinder Block, Materials for Cylinder Block with Justifications. Design of Bore Diameter; Bore Length and Thickness of Cylinder Wall

4.2 Function of Piston, Materials with Justification and Design of Piston and Piston Pin

4.3 Function of Connecting Rod, Materials with Justification and Design of Connecting Rod

4.4 Function of Rocker Arm, Materials with Justification and Design of Rocker Arm (For Rectangular Cross Section Only)

4.5 Function of Valve Spring, Materials with Justification and Design of Valve Spring

4.6 Function of Push Rod, Material with Justification and Design of Push Rod

About this Chapter

After reading this chapter, students will be able to :

- Select the relevant material with justification for the given engine component.
- Explain stepwise design procedure for the given engine component.
- Calculate dimensions of the given engine component from the given data.
- Draw proportionate diagram of the given engine component.

4.1 CYLINDER AND CYLINDER HEAD (W-16)

The function of cylinder is to retain the working fluid and to guide the piston. The cylinders are usually made of cast iron or cast steel because of high heat absorption capacity and low wear and tear due to friction.

4.1.1 Design Procedure of Cylinder and Cylinder Head (W-15)

1. Thickness of Cylinder Wall :

There are two types of stresses are produced by gas pressure inside the cylinder :

(a) Longitudinal stress.

(b) Circumferential stress.

Let, D_o = Outside diameter of the cylinder in mm

D = Inside diameter of the cylinder in mm

p = Maximum pressure inside the engine cylinder in N/mm^2

t = Thickness of cylinder wall in mm

and $1/m$ = Poisson's ratio, it is usually taken as 0.25

∴ Longitudinal stress,

$$\sigma_l = \frac{\text{Force}}{\text{Area}} = \frac{\frac{\pi}{4} \times D^2 \times p}{\frac{\pi}{4}\left[(D_o)^2 - D^2\right]} = \frac{D^2 \cdot p}{(D_o)^2 - D^2}$$

and circumferential stress is given by,

$$\sigma_c = \frac{\text{Force}}{\text{Area}} = \frac{D \times l \times p}{2t \times l} = \frac{D \times p}{2t}$$

... (where l is the length of cylinder and area is the projected area)

$\therefore$ Net longitudinal stress $= \sigma_l - \dfrac{\sigma_c}{m}$

Net circumferential stress $= \sigma_c - \dfrac{\sigma_l}{m}$

Thickness of cylinder wall (t) is obtained by using a thin cylindrical formula i.e.

$$\boxed{t = \frac{p \times D}{2\sigma_c} + C}$$

where,

p = Maximum pressure inside the cylinder in N/mm^2

D = Inside diameter of the cylinder or cylinder bore in mm

σ_c = Permissible circumferential or hoop stress for the cylinder material in MPa or N/mm^2

Value for,

σ_c = 35 MPa to 100 MPa depends upon the size and material of cylinder

C = Allowance for reboring

The thickness of the cylinder wall (t) may also be obtained from the following empirical relation i.e.

$$\boxed{t = 0.045\,D + 1.6 \text{ mm}}$$

Other relations are as follows :

(i)   Thickness of the dry liner = 0.03 D to 0.035 D

(ii)   Thickness of water jacket wall = 0.032 D + 1.6 m

(iii)   Water or space between outer cylinder wall and inner jacket wall

= 10 mm to 75 mm = 0.08 D + 6.5 mm

2. **Bore and Length of the Cylinder :**

We know, the indicated power produced inside the engine cylinder,

$$\boxed{\text{I.P.} = \frac{p_{im} \times l \times A \times nk}{60}} \text{ Watts} \qquad \qquad \text{... (4.1)}$$

where,

p_{im} = Indicated mean effective pressure

D = Cylinder bore in mm

A = Cross-sectional area of cylinder in mm^2

$\therefore$ $A = \dfrac{\pi}{4} \times D^2$

l = Length of stroke in mm

N = Speed of the engine in r.p.m.

and n = Number of working strokes per min

= N, for two stroke engine

= $\dfrac{N}{2}$, for four stroke engine

K = Number of cylinders

From equation (4.1), bore diameter 'D' and length of stroke (l) is determined.

Length of stroke is generally taken as 1.25 D to 2D.

Length of cylinder,   L = 1.15 $\times$ Length of stroke

$\therefore$ $\boxed{L = 1.15\,l}$

3. Cylinder Flange and Studs :

Diameter of studs or bolts may be obtained by equating the gas load due to the maximum pressure in the cylinder to the resisting force offered by all the studs or bolts.

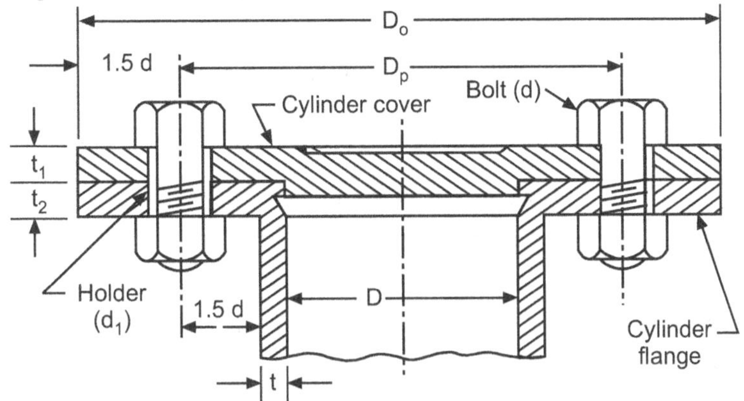

Fig. 4.1 : Design of cylinder head bolts or studs

$\therefore$

$$\frac{\pi}{4} \times D^2 \cdot p = n_s \times \frac{\pi}{4}(d_c)^2 \cdot \sigma_t$$

where,

D = Cylinder bore in mm

p = Maximum pressure in N/mm^2

n_s = Number of studs

$\therefore$ n_s = 0.01 D + 4 to 0.02 D + 4

d_c = Core or minor diameter i.e. diameter at the root of the thread in mm

σ_t = Allowable tensile stress in MPa or N/mm^2

d = Nominal or major diameter of the stud or bolt = 0.75 t_f to t_f

where, t_f = Thickness of flange

$d + 6$ = Distance of flange from the centre of the hole for the stud or bolt

4. Cylinder Head Thickness (th) :

The cylinder head may be approximately taken as a flat circular plate whose thickness (th) may be determined from the following relation :

$$t_h = D \sqrt{\frac{C.p.}{\sigma_c}}$$

where,

D = Cylinder bore in mm

p = Maximum pressure inside the cylinder in N/mm^2,

σ_c = Allowable circumferential stress in MPa or N/mm^2

It may be taken as 30 to 50 MPa

and C = Constant whose value is taken as 0.1

4.1.2 Indicated Power and Brake Power of an Engine Cylinder (W-18; S-16, 18)

Q.1. *Explain indicated power and brake power of an engine cylinder.*

Q.2. *Define indicated power and brake power.*

Indicated Power :   (S-16)

The power developed inside the cylinder is known as *indicated power*. It is called indicated power because it is measured from indicator diagram.

$$I.P. = \frac{p_{im} \times L \times A \times n \times k}{60} \text{ Watt}$$

where, p_{im} = Indicated mean effective pressure (N/mm^2)

L = Length of stroke in mm

A = Area of the piston in mm^2

N = Speed in revolutions per minute

n = Number of power strokes of engine

$n = \dfrac{N}{2}$ for four stroke engine

and   n = N for two stroke engine

k = Number of cylinders

Brake Power :   **(S-16)**

This is the actual power delivered at the crank shaft. It is obtained by deducting various power losses in the engine from indicated power. Brake power is what would keep the vehicle running at any speed once we have accelerated.

B.P. (in kW) can be calculated with the formula,

$$\boxed{\text{B.P.} = \frac{2\pi NT}{60}}\ \text{Watt}$$

where,   N = Engine speed in r.p.m.

T = Torque in Newton metres

OR $b_p = \dfrac{p_{im}\,LAnK}{60}$ Watt

where, p_{im} = Break mean effective pressure

Frictional Power :   **(S-16)**

Power lost in frictional losses at the working surfaces like bearing, piston rings, valves etc. is known as frictional power.

Relation between Indicated power, Brake power and Frictional power :

Frictional power = Indicated power – Brake power

SOLVED PROBLEMS

Problem 4.1 : *Calculate stroke length of four stroke engine with following specification :*

B.P. = 4.5 kW, Speed = 1200 rpm

Mean effective pressure (indicated) = 0.35 N/mm^2 and Mechanical efficiency = 80%.

Solution : Given :   B.P. = 4.5 kW = 4.5×10^3 Watt

N = 1200 rpm

p_{im} = 0.35 N/mm^2

η_{mech} = 80%

Let,   D = Bore of the cylinder in mm

A = Cross-sectional area of the cylinder = $\dfrac{\pi}{4} D^2$ mm^2

l = Length of stroke in mm = 1.5 D mm = $\dfrac{1.5\,D}{1000}$ m   ... Assume

We know that, $\text{I.P.} = \dfrac{\text{B.P.}}{\eta_{mech}} = \dfrac{4.5 \times 10^3}{0.8} = 5625$ Watt

∴   Indicated power $= \dfrac{p_{im} \cdot l \cdot A \cdot n}{60}$ Watt   where, $n = \dfrac{N}{2}$ ... for 4-stroke

$$\therefore \quad 5625 = \frac{0.35 \times 1.5\,D \times \frac{\pi}{4} D^2 \times \frac{1200}{2}}{60 \times 1000} = \frac{0.35 \times 1.5\,D^3 \times \pi \times 600}{4 \times 60 \times 1000}$$

$$\therefore \quad 5625 = \frac{989.1\,D^3}{4 \times 60 \times 1000} = 4.12\,D^3 \times 10^{-3}$$

$$\therefore \quad D^3 = \frac{5625}{4.12 \times 10^{-3}} = 1365.29 \times 10^3$$

$$\therefore \quad D = 110.94 \text{ mm} \quad \text{Say } D = 111 \text{ mm}$$

$\therefore$ Length of stroke, $l = 1.5\,D = 1.5 \times 111 = \mathbf{166.5}$

Problem 4.2 : *A four stroke diesel engine has the following specifications :*

$$Brake\ power\ =\ 5\ kW$$
$$Speed\ =\ 1200\ rpm$$

Indicated mean effective pressure = 0.35 N/mm^2

$$Mechanical\ efficiency\ =\ 80\%$$

Assume length of stroke, l = 1.5 D, or l = 1.08 D, C = 0.1

Tensile stress for cylinder cover = 52 N/mm^2.   **(S-15, 16, 17, 18)**

Determine dimensions of cylinder, thickness of cylinder head, size of studs for cylinder head.

Solution : Given :   B.P. $= 5$ kN $= 5 \times 10^3$ Watt

$$N = 1200 \text{ rpm} \left[\text{for 4-stroke; } n = \frac{N}{2} = 600 \right]$$

$$p_{im} = 0.35 \text{ N/mm}^2$$

$$\eta_m = 80\%$$

$$n = \frac{N}{2} = \frac{1200}{2} = 600, \quad A = \frac{\pi}{4}D^2 \text{ mm}^2$$

(a) Bore and length of cylinder :

Indicated power, $\text{I.P.} = \dfrac{\text{B.P.}}{\eta_m} = \dfrac{5000}{0.8} = 6250$ Watt

Length of stroke, $l = 1.5\,D \text{ mm} = \dfrac{1.5\,D}{1000} \text{ m}$

We know that, Indicated power (I.P.)

$$\text{I.P.} = \frac{p_{im}\,l\,A\,n}{60}$$

$$\therefore \quad 6250 = \frac{0.35 \times 1.5\,D \times \pi D^2 \times 600}{60 \times 1000 \times 4} = \frac{989.1\,D^3}{60000 \times 4}$$

$$\therefore \quad 989.1\,D^3 = 6250 \times 240 \times 10^3 = 150 \times 10^7$$

$$\therefore \quad D^3 = \frac{150 \times 10^7}{989.1} = 1516530.179$$

$$\therefore \quad D = \mathbf{114.89} \text{ say } \mathbf{115 \text{ mm}}$$

$\therefore$ Length of stroke, $l = 1.5\,D = 1.5 \times 115 = 172.5$ mm

Taking a clearance on both sides of the cylinder equal to 15% of the stroke, therefore length of the cylinder,

$$L = 1.15\,l = 1.15 \times 172.5 = 198.37 \text{ say}$$

$$\therefore \quad L = \mathbf{200 \text{ mm}}$$

(b) Thickness of the cylinder head :

The maximum pressure (p) in the engine cylinder is taken as 9 to 10 times the mean effective pressure (p_{im}).

$$\therefore \qquad p = 9\, p_{im} = 9 \times 0.35 = 3.15 \text{ N/mm}^2$$

Thickness of cylinder head,

$$t_h = D\sqrt{\frac{C.p}{\sigma_t}} = 115\sqrt{\frac{0.1 \times 3.15}{52}} \qquad \text{(Taking C = 0.1, } \sigma_t = 52 \text{ N/mm}^2)$$

$$\therefore \qquad t_h = 8.95 \text{ mm} \quad \text{say } \textbf{9 mm}$$

(c) Size of studs for the cylinder head :

Let,

$\quad d$ = Nominal diameter of the stud in mm

$\quad d_c$ = Core diameter of the stud in mm, it is usually taken as 0.84 d

$\quad \sigma_t$ = Tensile stress for the material of the stud which is usually nickel steel

$\quad \eta_s$ = Number of studs

We know that, the force acting on the cylinder head, (or on the studs)

$$= \frac{\pi}{4} \times D^2 \times p = \frac{\pi}{4} \times (115)^2 \times 3.15 = 32702 \text{ N} \qquad \ldots (1)$$

And number of studs, $\quad \eta_s = 0.01\, D + 4 = 0.01 \times 115 + 4 = 5.15$

or $\qquad\qquad\qquad\quad \eta_s = 0.02\, D + 4 = 0.02 \times 115 + 4 = 6$

Take $\qquad\qquad\qquad\quad \eta_s = \textbf{6}$

We know that, resisting force offered by all the studs

$$= \eta_s \times \frac{\pi}{4}\, (d_c)^2 \times \sigma_c = 6 \times \frac{\pi}{4} \times (0.84\, d)^2 \times 65$$

$$= 216\, d^2 \text{ N} \qquad \ldots (2) \text{ (Taking } \sigma_c = 65 \text{ MPa} = 65 \text{ N/mm}^2)$$

From equation (1) and (2), $d^2 = \dfrac{32702}{216}$

$$\therefore \qquad d = 151 \text{ or } d = \textbf{12.3 say 14 mm}$$

$$\text{P.C.D of the studs } (D_p) = D + 3d = 115 + 3 \times 14 = \textbf{157 mm}$$

$$\text{Pitch of the studs} = \frac{\pi \times D_p}{\eta_s} = \frac{\pi \times 157}{6} = 82.2 \text{ mm}$$

$$\text{Minimum pitch of studs} = 19\sqrt{d} = 19\sqrt{14} = \textbf{71.1 mm}$$

$$\text{Maximum pitch of studs} = 28.5\sqrt{d} = 28.5\sqrt{14} = \textbf{106.6 mm}$$

$\therefore \quad$ Pitch of stud lies between 71.1 mm to 106.6 mm.

$\therefore \quad$ Size of stud, d = 14 mm is safe.

Problem 4.3 : *Determine the thickness of plain cylinder head for 300 mm cylinder diameter. The maximum gas pressure is 3.2 N/mm². Take allowable tensile stress for cylinder cover is 42 N/mm² and constant is 0.1.*   **(S-17, W-17)**

Solution : Given Data : D = 300 mm, P_{max} = 3.2 N/mm², σ_t = 42 N/mm², C = 0.1

Thickness of plain cylinder head :

$$t = D\sqrt{\frac{C \times P_{max}}{\sigma_t}} = 300\sqrt{\frac{0.1 \times 3.2}{42}} = 26.18 \text{ mm} \cong 27 \text{ mm}$$

Problem 4.4 : *Determine the bore and length of cylinder of 4-stroke diesel engine for following specification - Brake power - 5 kW, Speed - 1200 r.p.m., P_m - 0.35 N/mm², Mechanical efficiency - 80%, L/D = 1.08.*   **(W-17)**

Solution : Given Data : B.P. = 5 kW = 5000 W

$$N = 1200 \text{ r.p.m. or } n = N/2 = 600$$
$$P_m = 0.35 \text{ N/mm}^2 = 0.35 \times 10^6 \text{ N/m}^2$$
$$\eta_m = 80\% = 0.8$$
$$L = 1.08 \, D$$

Bore and Length of Cylinder :

Let, D = Bore of the cylinder in mm

$$A = \text{Across section, Area of the cylinder} = \frac{\pi}{4} \times D^2 \text{ mm}^2$$

We know that, the indicated power,

$$\text{I.P.} = \frac{\text{B.P.}}{\eta_m} = \frac{5000}{0.8} = 6250 \text{ Watt}$$

We also know that, the indicated power (I.P.)

$$6250 = \frac{P_m \, l \cdot A \cdot n}{60} = \frac{0.35 \times 10^6 \times 1.08 \, D \times \pi D^2 \times 600}{60 \times 4} = 2.96 \times 10^6 \, D^3$$

$$\therefore \qquad D^3 = \frac{6250}{2.96 \times 10^6} = 2.11 \times 10^{-3} \text{ or } D = 0.128 \text{ m} = 128 \text{ mm}$$

$$L = 1.08 \times 128 = 138.24 \text{ mm} \cong 139 \text{ mm}$$

Taking a clearance on both sides of the cylinder equal to 15% of the stroke.

$$\therefore \qquad \text{Length of cylinder} = 1.15 \, L = 1.15 \times 139 = \textbf{160 mm}$$

Problem 4.5 : *A 4-stroke diesel engine has the following specifications :*
Brake power = 6 kW, Speed = 1200 r.p.m., Indicated mean effective pressure = 0.35 N/mm²,
Mechanical efficiency = 80%. Determine :
(a) Bore and length of cylinder,   (b) Thickness of cylinder head.   **(W-18)**

Solution : Assume l = 1.5 D or l = 1.08 D, Constant C = 0.1,
Tensile stress for cylinder cover = 52 N/mm².

Given Data :   B.P. = 6 kW = 6000 W

$$N = 1200 \text{ r.p.m.}$$

$$n = \frac{1200}{2} = 600 \text{ r.p.m.} \qquad\qquad \text{For four stroke engine}$$

$$P_m = 0.35 \text{ N/mm}^2$$
$$\eta_m = 80\% = 0.8$$

For cylinder cover, $\sigma_t = 52 \text{ N/mm}^2$ Assumed

Length of stroke, $L = 1.5 \, D = \dfrac{1.5 \, D}{1000} \text{ m}$ Assumed

1. Bore and length of cylinder :

Let D = Bore of cylinder in mm

A = Cross-sectional area of cylinder

$$= \frac{\pi}{4} D^2$$

We know that indicated power,

$$\text{I.P.} = \frac{\text{B.P.}}{\eta_m} = \frac{6000}{0.8} = 7500 \text{ W}$$

We also know that, $\text{I.P.} = \dfrac{P_m \, LAn}{60}$

$$\therefore \qquad 7500 = \frac{0.35 \times 1.5 \, D \times \pi D^2 \times 600}{60 \times 1000 \times 4} = 4.12 \times 10^{-3} \, D^3$$

$$\therefore \qquad D^3 = \frac{7500}{4.12 \times 10^{-3}} = 1818.91 \times 10^3$$

$$\therefore \qquad D = 122.06 \text{ mm}$$

$$\therefore \qquad D = \textbf{124 mm}$$

$$L = 1.5\,D = 1.5 \times 124 = 186 \text{ mm}$$

Taking a clearance on both sides of the cylinder equal to 15% of the stroke, therefore length of the cylinder,

$$\text{Length of cylinder} = 1.15 \times L = 1.15 \times 186 = 213.9 = \textbf{214 mm}$$

2. Thickness of the cylinder head : Since, the maximum pressure (P) in the engine cylinder is taken as 9 to 10 times means effective pressure (P_m) therefore, let us take,

$$P = 9P_m = 9 \times 0.35 = 3.15 \text{ N/mm}^2$$

We know that, thickness of the cylinder head,

$$t_h = D\sqrt{\frac{C \times P}{\sigma_t}} = 124 \times \sqrt{\frac{(0.1 \times 3.15)}{52}} = \textbf{9.65 mm} \quad \text{say } \textbf{10 mm}$$

Problem 4.6 : *Determine the thickness of plain cylinder head for 0.4 m cylinder diameter. The maximum gas pressure is 3.2 N/mm². Design the studs and cylinder cover. Take allowable tensile stress for cylinder cover and bolt equal to 42 N/mm² and 63 N/mm² respectively.* **(W-18)**

Solution : Given Data : $D = 0.4 \text{ m} = 400 \text{ mm},$

$$P = 3.2 \text{ N/mm}^2$$

$$C = 0.1 \qquad\qquad\qquad\qquad\qquad \text{Assumed}$$

$$\sigma_{t\,(cylinder)} = 42 \text{ N/mm}^2$$

$$\sigma_{t\,(bolt)} = 63 \text{ N/mm}^2$$

1. Thickness of plain cylinder :

$$t_h = D\sqrt{\frac{C \times P}{\sigma_{t\,(cylinder)}}} = 400 \times \sqrt{\frac{0.1 \times 3.2}{42}} = \textbf{34.91 mm} \quad \text{say } \textbf{35 mm}$$

2. Design of studs and cylinder cover :

Let, $d = $ Nominal diameter of stud

$$d_c = \text{Core diameter of stud (0.84 d)}$$

$$\sigma_{t\,(bolt)} = 63 \text{ N/mm}^2$$

$$n_s = \text{Number of studs}$$

We know that, the force acting on the cylinder head (or on the studs)

$$= \frac{\pi}{4} \times D^2 \times P = \frac{\pi}{4} \times 400^2 \times 3.2 = 402123.8 \text{ N}$$

The number of studs usually taken between

$n_s = 0.01\,D + 4 = (0.01 \times 400) + 4 = 8$ and $0.02\,D + 4 = (0.02 \times 400) + 4 = 12$. Taking $n_s = 12$

We know that, resisting force offered by all the studs

$$402123.8 = n_s \times \frac{\pi}{4}(d_c)^2 \times \sigma_{t\,(bolt)} = 12 \times \frac{\pi}{4}(0.84\,d)^2 \times 63$$

$$\therefore \qquad d = 28.39 \text{ mm} = \textbf{30 mm}$$

The pitch circle diameter of the stud (D_p) is taken as D + 3d

$$D_p = 400 + (3 \times 30) = 490 \text{ mm}$$

We know that, Pitch of the studs $= \dfrac{\pi D_p}{n_s} = \dfrac{\pi \times 490}{12} = 128.28 \text{ mm}$

For leak proof joint, the pitch of the stud should lie between $19\sqrt{d}$ to $28.5\sqrt{d}$, where d is nominal diameter of the stud.

$\therefore$ Minimum pitch of the stud = $19\sqrt{d} = 19 \times \sqrt{30} = 104.06$ mm

And Maximum pitch of the stud = $28.5\sqrt{d} = 28.5 \times \sqrt{30} = 156.10$ mm

Since, the pitch of the stud obtained above (i.e. 128.28 mm) lies between 104.06 mm and 156.10 mm, therefore size of the stud (d) calculated above is satisfactory.

$\therefore$ $\qquad\qquad\qquad\qquad$ d = **30 mm**

4.2 PISTON $\hfill$ (S-17)

Q.1. *What points are taken into consideration for design of the piston ?* $\hfill$ (S-15)

Q.2. *Draw a neat sketch of piston showing thrust and non-thrust side.* $\hfill$ (W-14)

The piston is a disc which reciprocates within a cylinder. The main function of the piston of an internal combustion engine is to receive the impulse from the expanding gas and to transmit the energy to the crankshaft through the connecting rod.

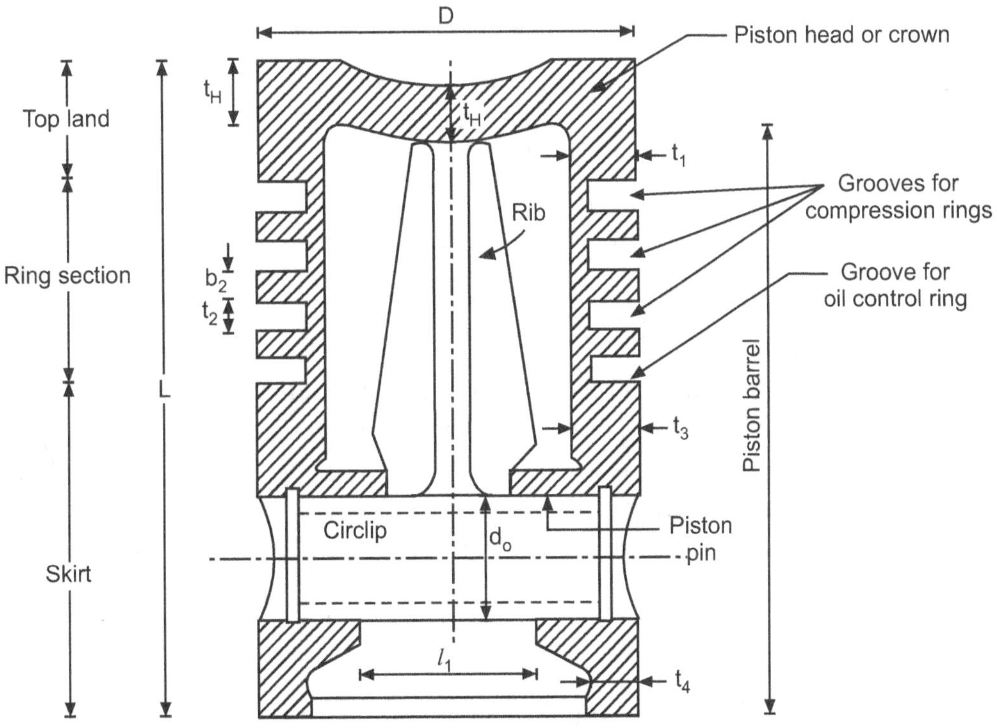

Fig. 4.2 : Design of piston

Design Considerations for Piston Material : $\hfill$ (S-16, 17)

1. It should withstand high gas pressure.
2. It should have to form an effective gas and oil sealing.
3. It should reciprocate at high speed without any noise.
4. It should withstand thermal and mechanical distortion.
5. It should have support for the piston pin.
6. It should possess minimum weight.
7. It should have minimum work friction.
8. It should have no corrosion or picking up.

Material for Pistons :

The most commonly used materials for pistons of I.C. engines are cast iron, cast aluminium, forged aluminium, cast steel and forged steel, because of high heat absorption capacity, shock resistance, low wear and tear of material.

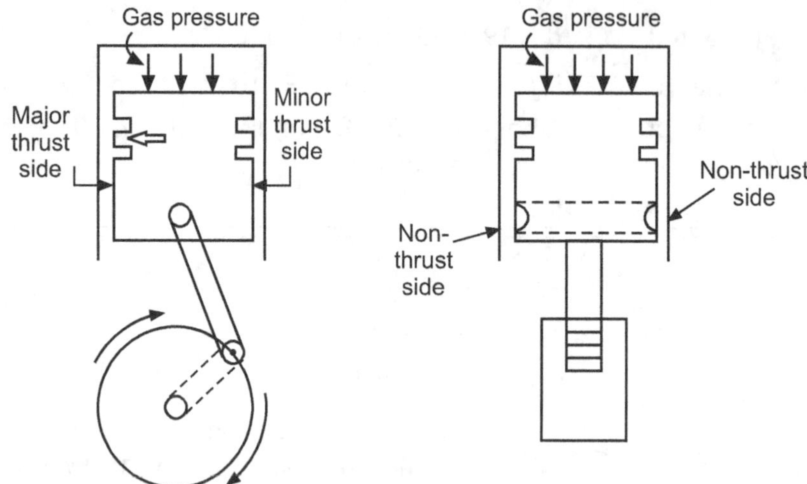

(a) Thrust side of I.C. engine piston   (b) Non-thrust side of I.C. engine piston (S-17, W-18)

Fig. 4.3

4.2.1 Design Procedure of Piston Head or Crown   (W-16, 17)

Q.1. *Explain design of piston crown for bending strength.*   (S-14)

1. Considering bending stress :

The thickness of the piston head (t_H), according to Grashoff's formula is given by,

$$t_h = \sqrt{\frac{3p \cdot D^2}{16\,\sigma_t}} \text{ in mm}$$

where,

p = Maximum gas pressure or explosion pressure in N/mm^2

D = Cylinder bore or outside diameter of the piston in mm

and

σ_b = Permissible bending stress for the material of the piston in MPa or N/mm^2

2. Considering thermal or heat transfer :

The heat absorbed by the piston due to combustion of fuel is quickly transfered to the walls of the cylinder.

Considering piston head as a circular plate, its thickness is given by,

$$t_h = \frac{H}{12.56\,k\,(T_C - T_E)} \text{ in mm}$$

where,

H = Heat flowing through the piston head in kJ/s

k = Heat conductivity factor in W/m/°C

For grey cast iron, k = 46.6 W/m/°C

For steel, k = 51.25 W/m/°C

For aluminium alloys, k = 174.75 W/m/°C

T_C = Temperature at the centre of the piston head in °C

and

T_E = Temperature at the edges of the piston head in °C

Heat flowing through piston head (H) is determined by following expression :

$$H = C \times HCV \times m \times B.P. \text{ in kW}$$

where,

C = Constant = 0.05

HCV = Higher calorific value of the fuel in kJ/kg

It may be taken as 45×10^3 kJ/kg for diesel and 47×10^3 kJ/kg for petrol

m = Mass of the fuel used in kg per brake power per second

and

$B.P.$ = Brake power of the engine per cylinder

SOLVED PROBLEMS

Problem 4.7 : *Design the piston crown thickness from the following data :*

Diameter of piston = 80 mm, Maximum pressure on the piston = 4.5 N/mm^2

and Allowable bending stress = 45 N/mm^2

Solution : Given : $D = 80$ mm, $P = 4.5$ N/mm^2, $\sigma_b = 45$ N/mm^2

Let the thickness of piston head can be designed by assuming the head to be a flat plate uniform thickness and fixed at the edges and assuming the gas load to be uniform distributed.

$$t_h = \sqrt{\frac{3p \cdot D^2}{16\,\sigma_b}}$$

where, t_h = Thickness of piston crown

$p = p_{max}$

$\therefore$ $t_h = \sqrt{\dfrac{3 \times 4.5 \times 80^2}{16 \times 45}}$ = **10.95 mm**   say **11 mm**

Problem 4.8 : *Design the piston crown thickness from the following data : diameter of piston = 90 mm, maximum pressure on the piston = 4.6 N/mm^2 and allowable bending stress = 45 N/mm^2.*   **(W-18)**

Solution : Given Data : $D = 90$ mm, $P = 4.6$ N/mm^2, $\sigma_b = 45$ N/mm^2.

Let the thickness of the piston head can be designed by assuming the head to be a flat plate uniform thickness and fixed at the edges and assuming the gas load to be uniformly distributed.

$$t_h = \sqrt{\frac{3\,PD^2}{16\,\sigma_b}}$$

where, t_h = Thickness of piston crown

$P = P_{max}$

$\therefore$ $t_h = \sqrt{\dfrac{3 \times 4.6 \times 90^2}{16 \times 45}}$ = **12.32 mm**   say **13 mm**

Problem 4.9 : *Design an aluminium alloy piston crown on the basis of strength and heat dissipation for a single acting four stroke engine for the following specifications :*

Cylinder bore = 300 mm,	*Speed = 500 rpm,*
Stroke = 375 mm,	*Bending stress in piston crown = 36 MPa,*
Maximum gas pressure = 8 MPa,	*Crown temperature difference = 70°C,*
Brake mean effective pressure = 1.15 MPa,	*Heat conductivity = 160 W/m/°C,*
Fuel consumption = 0.22 kg/kW/hr,	*Mechanical efficiency = 80%.*

Solution : Given :

$D = 300$ mm

$L = 375$ mm $= 0.375$ m

$p_m = 1.15$ MPa $= 1.15$ N/mm^2

$p = 8$ N/mm^2

m = Fuel consumption $= 0.22$ kg/kW/hr

$= 61.11 \times 10^{-6}$ kg/B.P./sec

$N = 500$ rpm

$\sigma_b = 36$ MPa $= 36$ N/mm^2

$T_C - T_E = 70$°C

$K = 160$ W/m°C

Thickness of piston head or crown :

(a) On the basis of strength :

$$t_h = \sqrt{\frac{3p \cdot D^2}{16\,\sigma_b}} = \sqrt{\frac{3 \times 8 \times (300)^2}{16 \times 36}} = \sqrt{\frac{2160000}{576}} = \sqrt{3750}$$

$\therefore \qquad t_h = \textbf{61.23 mm} \ \ \text{say } \textbf{62 mm}$

(b) On the basis of heat dissipation :

Since, engine is a four stroke engine.

$\therefore$ Number of working strokes per minute,

$$n = \frac{N}{2} = \frac{500}{2} = 250$$

Cross-sectional area of the cylinder,

$$A = \frac{\pi}{4}D^2 = \frac{\pi \times (300)^2}{4} = 70{,}650 \ \text{mm}^2$$

Indicated power, $\qquad$ $\text{I.P.} = \dfrac{p_m\, L \cdot A \cdot n}{60} = \dfrac{1.15 \times 0.375 \times 70650 \times 250}{60}$

$\therefore \qquad \text{I.P.} = 126949.22 \ \text{Watt}$

$\therefore \qquad \text{I.P.} = 126.95 \ \text{kW}$

$\therefore \qquad \text{B.P.} = \text{I.P.} \times \eta_m = 126.95 \times 0.8$

$\qquad \text{B.P.} = 101.56 \ \text{kW}$

We know that heat flowing through the piston head,

$$H = C \times HCV \times m \times B.P. = 0.05 \times 42 \times 61.11 \times 10^{-6} \times 101.56$$
$$= 0.0130 \ \text{kW} = \textbf{13.033 Watt} \ (\text{taking } C = 0.05)$$

$\therefore$ Thickness of the piston head,

$$t_h = \frac{H}{12.56\, K\,(T_C - T_E)} = \frac{13.033}{12.56 \times 160 \times 70} = \frac{13.033}{140672}$$

$\therefore \qquad t_h = 9.264 \times 10^{-5} = \textbf{0.0926 mm}$

4.2.2 Design of Piston Rings $\qquad\qquad$ (S-16, 18; W-16)

Q.1. *Write design calculation for piston rings.* $\qquad\qquad$ **(W-14)**

(a) The radial thickness (t_1) of the ring may be obtained by considering the radial pressure between the cylinder wall and the ring. From bending stress consideration in the ring, the radial thickness is given by,

$$\boxed{\,t_1 = D\sqrt{\frac{3\,p_w}{\sigma_b}}\,} \ \ \text{in mm}$$

where, D = Cylinder bore in mm

$\qquad\quad$ p_w = Pressure of gas on the cylinder wall in N/mm^2

$\qquad\quad$ σ_b = Allowable bending (tensile) stress in MPa

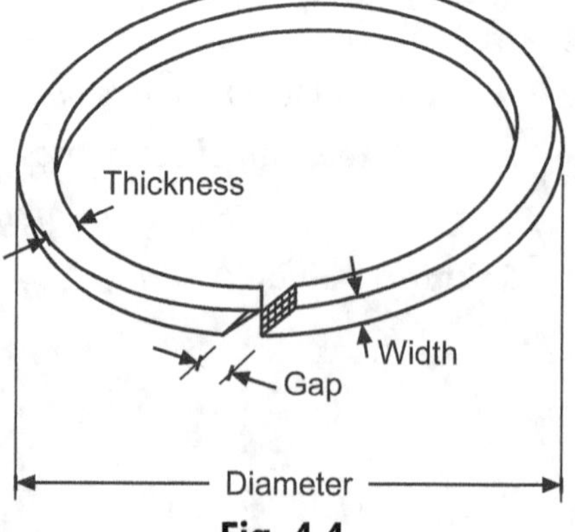

Fig. 4.4

(b) Axial thickness (t_2) of the rings may be taken as 0.7 t_1 to t_1.

The maximum axial thickness (t_2) may also be obtained from the following empirical relation :

$$\boxed{\,t_2 = \frac{D}{10\,n_R}\,}$$

where, $\qquad\qquad$ n_R = Number of rings

(c) Width of top land,
$$b_1 = t_H \text{ to } 1.2 \, t_H$$
(d) Width of other ring lands,
$$b_2 = 0.75 \, t_2 \text{ to } t_2$$
(e) The gap between the free ends of the ring is given by, $3.5 \, t_1$ to $4 \, t_1$.

SOLVED PROBLEMS

Problem 4.10 : *If radial width of piston ring is 8 mm and if two compression rings and one oil ring are to be employed on piston having crown thickness of 10 mm, compute length of piston above skirt.* **(W-15)**

Solution : Given : Compression rings = 2,

$$\text{Oil ring} = 1$$

Width of piston ring, $\quad b_r = 8$ mm

Crown thickness, $\quad t_h = 10$ mm

1. **Top land (b_1) :** $\quad b_1 = t_h \text{ to } 1.2 \, t_h = 10 \text{ to } 10 \times 1.2 = 10 \text{ to } 12$ mm

Consider, Top land (b_1) = 11 mm

2. **Length of the ring section :**

The axial thickness (t_2) of the rings = $0.7 \, b_r = 0.7 \times 8 = 5.6$ mm

Length of other land (b_2) = $0.85 \, b_r = 0.85 \times 8 = 6.8$ mm

Length of the ring section = $3 \, t_2 + 2 \, b_2 = 3 \times 5.6 + 2 \times 6.8 = 30.4$ mm

Length of piston above skirt = Top land (b_1) + Length of the ring section.

∴ Length of piston above skirt = 11 + 30.4 = **41.4 mm.**

Problem 4.11 : *Design piston ring for following : Number of ring = 5, Wall pressure (P_w) = 0.035 N/mm^2, Bending stress for ring = 85 N/mm^2, Diameter of cylinder bore = 240 mm.* **(W-17)**

Solution : Given data : Number of ring, $n_R = 5$, $D = 240$ mm, $P_w = 0.035$ N/mm^2, $\sigma_t = 85$ N/mm^2

The radial thickness (t_1) of the ring may be obtained by considering the radial pressure between the cylinder wall and the ring. From bending stress consideration in the ring, the radial thickness is given by,

$$t_1 = D\sqrt{\frac{3 p_w}{\sigma_t}} = 240 \sqrt{\frac{(3 \times 0.035)}{85}} = 8.43 \text{ mm} = \text{Say 9 mm}$$

where, $\qquad D$ = Cylinder bore in mm,

$\qquad p_w$ = Pressure of gas on the cylinder wall in N/mm^2.

$\qquad \sigma_t$ = Allowable bending (tensile) stress in MPa.

The axial thickness (t_2) of the rings may be taken as $0.7 \, t_1$ to $t_1 = 0.7 \times 9 = 6.3$ mm.

The minimum axial thickness (t_2) may also be obtained from the following empirical relation :

$$t_2 = \frac{D}{10 \, n_R}$$

where, $\qquad n_R$ = Number of rings

$$t_2 = \frac{240}{(10 \times 5)} = 4.8 \text{ mm}$$

Selecting maximum value of above $t_2 = 6.3$ mm

Width of other ring lands, $b_2 = 0.75 \, t_2 \text{ to } t_2 = 0.85 \, t_2 = 0.85 \times 6.3 = 5.35$ mm

The gap between the free ends of the ring is given by $3.5 \, t_1$ to $4 \, t_1$

$$= 3.75 \, t_1 = 3.75 \times 9 = 33.75 \text{ mm}$$

Length of the ring section :

Length of ring section = $5 t_2 + 4 b_2 = (5 \times 6.3) + (4 \times 5.35) = $ **52.90 mm**

4.2.3 Design of Piston Skirt   (S-16, 18; W-16)

The portion of the piston below the ring section is known as *piston skirt*.

It acts as a bearing for the side thrust of the connecting rod.

We know that, maximum gas load on the piston,

$$P = p \times \frac{\pi}{4} D^2$$

∴ Maximum side thrust on the cylinder,

$$R = \frac{P}{10} = 0.1\, p \times \frac{\pi}{4} D^2 \qquad \qquad \text{... (4.2)}$$

where, p = Maximum gas pressure in N/mm^2

and D = Cylinder bore in mm

The side thrust (R) is also given by,

$$R = [\text{Bearing pressure}] \times [\text{Projected bearing area of the piston skirt}]$$
$$= p_b \times D \times l \qquad \qquad \text{... (4.3)}$$

where, l = Length of the piston skirt in mm

From equation (4.2) and (4.3), the length of the piston skirt (l) is determined.

Generally, the length of the piston skirt is taken as 0.65 to 0.8 times the cylinder bore.

∴ Total length of the piston (L) is given by

$$\boxed{L = \text{Length of the skirt + Length of ring section + Top land}}$$

SOLVED PROBLEMS

Problem 4.12 : *Design the skirt length of the piston with given data for petrol engine. Maximum pressure inside the cylinder = 4.5 N/mm^2. Piston diameter = 70 mm. Side thrust is limited to 8% of the maximum load on the piston. Allowable bearing pressure = 0.3 N/mm^2, also draw a neat sketch.*   **(S-15)**

Solution : Given : p_{max} = 4.5 N/mm^2

Piston diameter, D = 70 mm

$$\text{Side thrust} = 8\% = \frac{8}{100} = 0.08$$

p_b = 0.3 N/mm^2

Let, R = Normal side thrust acting on piston skirts

F = Total force produced due to combustion

p_{max} = Maximum gas pressure inside the cylinder

D = Diameter of piston

$$F = p_{max} \times \frac{\pi}{4} D^2 = 4.5 \times \frac{\pi}{4} (70)^2$$

∴ $F = 17.318 \times 10^3$ N

∴ $R = 0.08 \times F$

∴ $R = 0.08 \times 17.318 \times 10^3$ (∵ Side thrust = 8%)

∴ R = **1385.44 N**

Let, l_1 = Length of the piston skirt

The piston skirt acts as a bearing inside the liner.

We have, $\boxed{R = l_1 \times D \times p_b}$

where, p_b = Allowable bearing pressure on the piston skirt

$\therefore$ $l_1 = \dfrac{1385.44}{70 \times 0.3}$

$\therefore$ Length of piston skirt, l_1 = 65.97 mm = **66 mm**

Problem 4.13 : *Design the skirt length of the piston with the following data of petrol engine. Maximum pressure inside the cylinder = 6.5 N/mm². Piston diameter = 100 mm, side thrust is limited to 10% of maximum load on the piston. Allowable bearing pressure = 0.3 N/mm².*   **(W-17)**

Solution : Given data : P_{max} = 6.5 N/mm²

Piston diameter = D = 100 mm

Side thrust = 10% = 0.10

P_b = 0.3 N/mm²

Let,   R = Normal side thrust acting on piston skirts

F = Total force produced due to combustion

P_{max} = Maximum gas pressure inside the engine

D = Diameter of piston

$$F = P_{max} \times \frac{\pi}{4} D^2 = 6.5 \times 0.786 \times 100^2 = 51050.8 \text{ N}$$

R = 0.1 F   Side thrust = 10% = 0.10

$\therefore$ R = 5105.08 N

Let, l_1 = Length of piston skirt

The piston skirt acts as a bearing inside the liner.

We have,   R = $l_1 \times D \times P_b$

$\therefore$ 5105.08 = $l_1 \times 100 \times 0.3$

Where, P_b = Allowable bearing pressure on the piston skirt

$\therefore$ $l_1 = \dfrac{5105.08}{(100 \times 0.3)}$ = **170.16 mm**

4.2.4 Design of Piston Pin for Bearing, Bending and Shear Consideration

(W-15)

1.   Design of piston pin on the basis of bearing pressure :

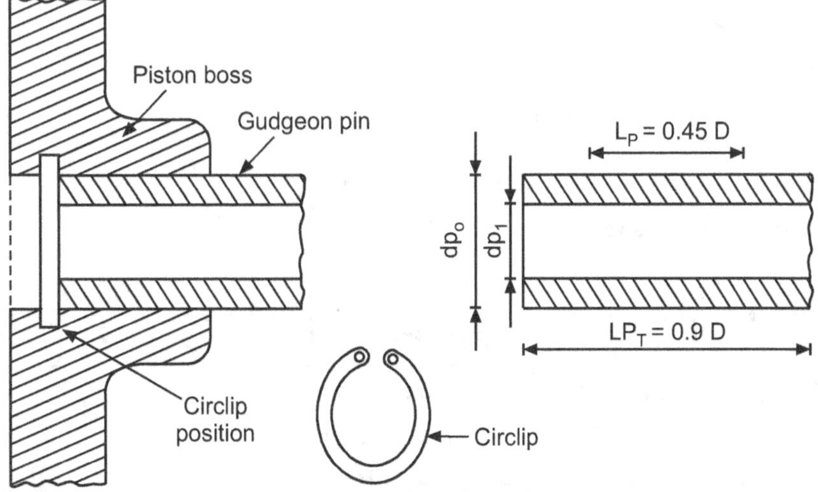

Fig. 4.5 : Piston pin

Let, $\qquad$ d_o = Outside diameter of the piston pin in mm

$\qquad$ l_p = Length of the piston pin in the bush of the small end of the connecting rod in mm. Its value is taken as 0.45 D

$\qquad$ p_{b1} = Bearing pressure at the small end of the connecting rod bushing in N/mm^2. Its value for the bronze bushing may be taken as 25 N/mm^2

We know that, load on the piston due to gas pressure or gas load,

i.e. $\qquad$ Maximum gas load $= \dfrac{\pi D^2}{4} \times p_{max}$ $\hfill \dots (4.4)$

where, $\qquad$ D = Diameter of piston

and load on the piston pin due to bearing pressure or bearing load

$\qquad$ = Bearing pressure $\times$ Bearing area

$\therefore$ $\qquad$ Maximum gas load $= p_{b1} \times d_o \times l_p$ $\hfill \dots (4.5)$

From equations (4.4) and (4.5), the outside diameter of the piston (d_o) may be obtained.

2. **Design of piston pin on the basis of bending :**

Bending moment 'M' is calculated as,

$$M = \frac{\text{Maximum gas load (F)} \times \text{Diameter of piston}}{8}$$

$\therefore$ $\qquad$ $\boxed{M = \dfrac{F \times D}{8}}$ $\hfill \dots (4.6)$

Also, $\qquad$ $M = \dfrac{\pi}{32} \times \sigma_b \times (dp_o)^3$ $\hfill \dots (4.7)$

Equating equation (4.6) and (4.7), we get,

$$\boxed{\frac{F \times D}{8} = \frac{\pi}{32} \times \sigma_b \times (dp_o)^3}$$

where, $\qquad$ σ_b = Bending stress in N/mm^2

$\qquad$ dp_o = Outside diameter of pin in mm

From above equation 'dp_o' may be calculated.

3. **Design of piston pin on the basis of shear strength :**

   (i) **For solid pin :**

$$\text{Gas load} = 2 \times \frac{\pi}{4} d_o^2 \times \tau \qquad\qquad \dots \text{Considering double shear}$$

$\qquad$ From above equation 'd_o' can be calculated.

   (ii) **For hollow pin :**

$$\text{Gas load} = 2 \times \frac{\pi}{4} (d_o^2 - d_i^2) \times \tau \qquad\qquad \dots \text{Considering double shear}$$

From above equation d_o and d_i for pin can be calculated.

4. **Total length of piston pin is taken as :**

$$\boxed{L_{pt} = 0.9\,D}$$

SOLVED PROBLEMS

Problem 4.14 : *Design the piston pin with following data :*

$\qquad$ *Maximum gas pressure* $= 4\,N/mm^2$, *Diameter of piston* $= 70\,mm$ $\qquad$ **(W-16, 17; S-17, 18)**

Allowable stresses due to bearing, bending and shear are 30 N/mm^2, 80 N/mm^2, 60 N/mm^2 respectively.

Solution : Given : Diameter of piston = D = 70 mm, Maximum pressure = p_{max} = 4 N/mm^2,

Bearing pressure, $p_b = 30$ N/mm^2, Bending stress, $\sigma_b = 80$ N/mm^2, Shearing stress, $t = 60$ N/mm^2

Maximum gas load, $\qquad\qquad F = \dfrac{\pi}{4} \times D^2 \times p_{max} = \dfrac{\pi}{4} \times (70)^2 \times 4 = 15.3938 \times 10^3$ N

(a) Design the piston pin on the basis of bearing pressure :

Let, $\qquad\qquad\qquad\qquad dp_o =$ Outer diameter of piston pin

$\qquad\qquad\qquad\qquad l_{pt} =$ Length of piston pin

$\qquad\qquad\qquad\qquad\quad = 0.45\ D = 0.45 \times 70 = 31.5$ mm

$\qquad\qquad\qquad\qquad F = dp_o \times l_{pt} \times p_b$

$\therefore \qquad\qquad\qquad\qquad dp_o = \dfrac{F}{l_{pt} \times p_b}$

$\therefore \qquad\qquad\qquad\qquad dp_o = \dfrac{15.3938 \times 10^3}{31.5 \times 30} = 16.29$ mm say 17 mm

$\therefore \qquad$ Outside diameter of pin $= $ **17 mm**

(b) Designing the piston pin on the basis of bending :

The bending moment 'M' is calculated as,

$$M = \frac{F \times D}{8} = \frac{15.3938 \times 10^3 \times 70}{8} \text{ N-mm} = 134.69 \times 10^3 \text{ N-mm}$$

Also, $\qquad\qquad\qquad\qquad M = \dfrac{\pi}{32} \times \sigma_b \times (dp_o)^3$

$\therefore \qquad\qquad\qquad\qquad \sigma_b = $ **279.2589 N/mm^2**

$\therefore$ The induced bending stress is greater than the permissible bending stress. So redesign of pin is necessary.

$\therefore$ For, $\qquad\qquad\qquad\qquad \sigma_b = 80$ N/mm^2

$\therefore \qquad\qquad\qquad\qquad M = \dfrac{\pi}{32} \times \sigma_b \times (dp_o)^3$

$\therefore \qquad\qquad 134.69 \times 10^3 = \dfrac{\pi}{32} \times 80 \times (dp_o)^3$

$\therefore \qquad\qquad\qquad\qquad (dp_o)^3 = \dfrac{134.69 \times 10^3}{\dfrac{\pi}{32} \times 80}$

$\therefore \qquad\qquad\qquad\qquad (dp_o)^3 = \dfrac{134.69 \times 10^3}{7.85}$

$\therefore \qquad\qquad\qquad\qquad (dp_o)^3 = 17157.961$

$\therefore \qquad\qquad\qquad\qquad dp_o = $ **25.79 mm** Say **26 mm**

(c) Designing piston pin on the basis of shear stress due to double shear :

$$\text{Gas pressure } (F) = 2 \times \frac{\pi}{4} (Dp_o)^2 \times \tau$$

$\therefore \qquad\qquad 15.39 \times 10^3 = 2 \times \dfrac{\pi}{4} \times (26)^2 \times \tau$

$\therefore \qquad\qquad\qquad\qquad \tau = $ **14.49 N/mm^2**

$\therefore$ The induced shear stress is less than permissible shear stress, so the design is safe.

(d) The total length of piston pin is taken as :

$\qquad\qquad l_{pt} = 0.9\ D = 0.9 \times 70 = $ **63 mm**

Problem 4.15 : *Design a piston pin for a piston having diameter 100 mm and sustaining maximum gas pressure of 5 N/mm². The allowable stresses for pin material are 25 N/mm² in bearing, 70 N/mm² in shear and 140 N/mm² in bending. Draw a neat sketch of piston and locate piston pin centre on it.*

Assume maximum bearing pressure on piston is limited to 0.45 N/mm².   **(W-15)**

Solution : Given : Diameter of piston, D = 100 mm

Maximum gas pressure, p_{max} = 5 N/mm²

Bearing pressure, p_b = 25 N/mm²

Bending stress, σ_b = 140 N/mm²

Shearing stress, τ = 70 N/mm²

Let, R = Normal side thrust acting on piston skirts

Maximum gas load, $F = p_{max} \times \dfrac{\pi}{4} D^2 = 5 \times \dfrac{\pi}{4}(100)^2 = 39.2699 \times 10^3$ N

and $R = 0.1 \times F$ (Assume, Side thrust = 10%)

$\therefore$ $R = 0.08 \times 39.2699 \times 10^3 = 3926.9$ N

Let, l_1 = Length of piston skirt

The piston skirt act as a bearing inside the liner

We have, $R = l_1 \times D \times p_b$

where, p_b = Allowable bearing pressure on the piston skirt

$\therefore$ $l_1 = \dfrac{3926.9}{100 \times 0.45} = 87.264$ mm = 88 mm

1. Design of piston spring on the basis of bearing pressure :

Maximum gas pressure = Bearing load, $F = d_{po} \times l_p \times p_b$

where, d_{po} = Outer diameter of piston pin

l_p = Length of piston pin at small end of connecting rod

     $= 0.45 \times D = 0.45 \times 100 = 45$ mm

$F = d_{po} \times l_p \times p_b$

$\therefore$ $39.2699 \times 10^3 = d_{po} \times 45 \times 25$

$\therefore$ d_{po} = **34.90 $\cong$ 35 mm**

2. Design of piston pin on the basis of bending :

$$M = \frac{F \times D}{8} = \frac{39.2699 \times 10^3 \times 100}{8}$$

$\therefore$ $M = 490.862 \times 10^3$ N-mm

and $M = \dfrac{\pi}{32} \times \sigma_b \times d_{po}^3$

$\therefore$ $490.862 \times 10^3 = \dfrac{\pi}{32} \times \sigma_b \times 35^3$

$\sigma_b = 116.61$ N/mm²

The induced bending stress is less than permissible stress i.e. 140 N/mm², hence design is ok.

3. Design of piston pin on the basis of shear stress, due to double shear :

$$F = 2\frac{\pi}{4} \times d_{po}^2 \times \tau$$

$\therefore$ $39.2699 \times 10^3 = 2\dfrac{\pi}{4} \times 35^2 \times \tau$

$\therefore$ $\tau = 20.41$ N/mm²

The induced shear stress is less than permissible stress i.e. 70 N/mm², hence design is safe.

4. **Total length of piston pin :**

$$l_{pr} = 0.9\,D = 0.9 \times 100 = 90 \text{ mm}$$

5. **Location of piston pin center** $= \dfrac{l_1}{2} + 0.03\,D$ from bottom edge of piston

Here, $\qquad l_1 =$ Length of piston skirt $= 88$ mm

$\therefore$ Location of piston pin center $= \dfrac{88}{2} + 0.03 \times 100 = 47$ from bottom edge of piston

For Fig. : Refer Fig. 4.2.

Problem 4.16 : *Design piston pin for following data: piston diameter 70 mm, maximum gas pressure inside cylinder 4.5 N/mm². Allowable stresses in bending, shear and bearing are 100 MPa, 70 MPa and 25 MPa respectively.* **(S-16)**

Solution : Given : Diameter of piston, D = 70 mm

Maximum gas pressure, $\qquad p_{max} = 4.5 \text{ N/mm}^2$

Bending stress, $\qquad \sigma_b = 100 \text{ N/mm}^2$

Shearing stress, $\qquad \tau = 70 \text{ N/mm}^2$

Bearing pressure, $\qquad p_b = 25 \text{ N/mm}^2$

Let, $\qquad R =$ Normal side thrust acting on piston skirts

Maximum gas load, $\qquad F = p_{max} \times \dfrac{\pi}{4} D^2 = 4.5 \times \dfrac{\pi}{4}(70)^2 = 17.318 \times 10^3 \text{ N}$

1. **Design of piston pin on the basis of bearing pressure :**

Maximum gas pressure $=$ Bearing load, $F = d_{po} \times l_{pt} \times p_b$

where, $\qquad d_{po} =$ Outer diameter of piston pin

$\qquad l_p =$ Length of piston pin at small end of connecting rod

$\qquad = 0.45 \times D = 0.45 \times 70 = 31.5 \text{ mm}$

Now, $\qquad F = d_{po} \times l_{pt} \times p_b$

$\therefore \qquad 17.318 \times 10^3 = d_{po} \times 31.5 \times 25$

$\therefore \qquad d_{po} = \mathbf{21.99 \cong 22 \text{ mm}}$

2. **Design of piston pin on the basis of bending :**

$$M = \dfrac{F \times D}{8} = \dfrac{17.318 \times 10^3 \times 70}{8} = 151.53 \times 10^3 \text{ N-mm}$$

$$M = \dfrac{\pi}{32} \times \sigma_b \times d_{po}^3$$

$\therefore \qquad 151.53 \times 10^3 = \dfrac{\pi}{32} \times \sigma_b \times 22^3$

$\therefore \qquad \sigma_b = 144.9 \text{ N/mm}^2$

The induced bending stresses are greater than permissible stress i.e. 100 N/mm², hence redesign is necessary.

$$M = \dfrac{\pi}{32} \times \sigma_b \times d_{po}^3$$

$\therefore \qquad 151.53 \times 10^3 = \dfrac{\pi}{32} \times 100 \times d_{po}^3$

$\therefore \qquad d_{po} = \mathbf{24.89 \cong 25 \text{ mm}}$

3. **Design of piston pin on the basis of shear stress, due to double shear :**

$$F = 2\dfrac{\pi}{4} \times d_{po}^2 \times \tau$$

$$\therefore \qquad 151.53 \times 10^3 = 2\frac{\pi}{4} \times 25^2 \times \tau$$

$$\therefore \qquad \tau = \textbf{17.63 N/mm}^2$$

The induced shear stress is less than permissible stress i.e. 70 N/mm^2, hence design is safe.

4. Total length of piston pin :

$$l_{pr} = 0.9\,D = 0.9 \times 70 = \textbf{63 mm}$$

4.3 DESIGN OF CONNECTING ROD   (S-15)

> **Q.1.** *Draw a neat proportionate sketch of connecting rod. Why I-section is used as cross-section of it ? What material is selected for connecting rod ?*   **(S-15)**
>
> **Q.2.** *Why I-section is used as cross-section of connecting rod ?*

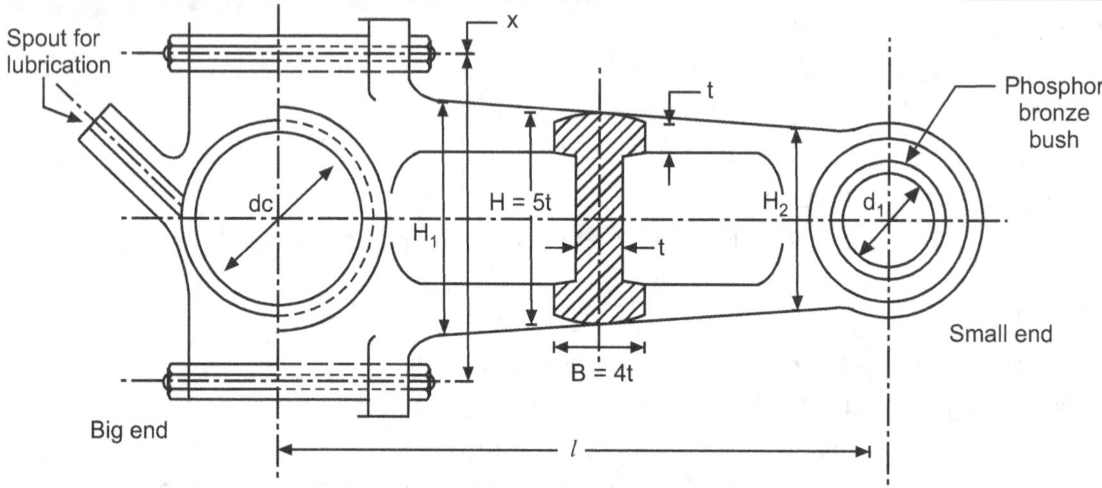

Fig. 4.6 : Design of connecting rod

I-sections are usually found to be most suitable for high speed engine connecting rod. Lightness is essential in order to keep inertia forces as small as possible. I-section also provides sufficient strength required to withstand momentary high gas pressure in the cylinder. I-section is four times stronger for buckling about X-X axis than Y-Y axis. Thus, I-section fulfils most desirable conditions for connecting rod i.e. adequate strength and stiffness and minimum weight.

4.3.1 Design Procedure of Connecting Rod   (W-16, 17; S-17)

> **Q.1.** *Explain design procedure of a connecting rod.*   **(W-14)**

(a) Dimensions of cross-section of the connecting rod :

According to Rankine's formula,

Buckling load,

$$W_B = \frac{\sigma_c \cdot A}{1 + a\left(\dfrac{L}{K_{XX}}\right)^2}$$

where,

A = Cross-sectional area of the connecting rod

$\therefore \qquad A = 11\,t^2$

L = Effective length of the connecting rod

σ_c = Crippling or Buckling stress

W_B = Buckling load

α = Rankine's constant

$K_{XX}^2 = 3.18\,t^2$

From this relation t (thickness of the flange and web of the section) can be determined.

The most suitable section for the connecting rod is I-section with the proportions as shown in Fig. 4.7.

Width of the section , B = 4t

and Depth or Height of the section, H = 5t

The dimensions B = 4t and H = 5t, as obtained above by applying the Rankine's formula, are at the middle of the connecting rod.

The width of the section (B) is kept constant throughout the length of the connecting rod, but the depth or height varies.

The depth near the small end (or piston end) is taken as $H_1 = 0.75\ H$ to $0.9\ H$.

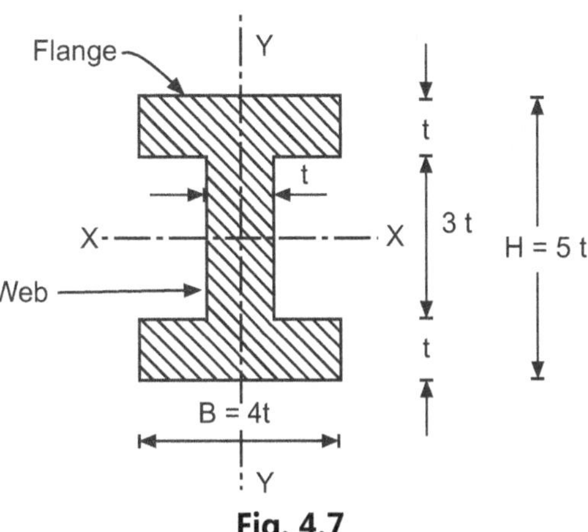

Fig. 4.7

The depth near the big end (or crank end) is taken as $H_2 = 1.1\ H$ to $1.25\ H$.

(b) Dimensions of the big end and small end of connecting rod :

Maximum gas force,

$$\boxed{F_{max} = \frac{\pi}{4} D^2 \times p}$$

... (4.8)

where,   D = Cylinder bore or piston diameter in mm

and   p = Maximum gas pressure in N/mm^2

Let, d_c = Diameter of the crank pin in mm

l_c = Length of the crank pin in mm

p_{bc} = Allowable bearing pressure in N/mm^2

and d_p, l_p and p_{bp} = Corresponding values for the piston pin

Load on the crank pin = Projected area × Bearing pressure

$= d_c \cdot l_c \cdot p_{bc}$

... (4.9)

Similarly, load on the piston pin $= d_p \cdot l_p \cdot p_{bp}$

... (4.10)

Equating equations (4.8) and (4.9), we have

$$\boxed{F_{max} = d_c \cdot l_c \cdot p_{bc}}$$

Taking $l_c = 1.25\ d_c$ to $1.5\ d_c$, the values of d_c and l_c are determined from the above expression.

Again, equating equation (4.8) and (4.10), we have

$$\boxed{F_{max} = d_p \cdot l_p \cdot p_{bp}}$$

Taking, $l_p = 1.5\ d_p$ to $2\ d_p$, the value of d_p and l_p are determined from the above expression.

(c) Size of bolts for securing the big end cap :

F_t = Inertia load acting on bolts

Let, dc_b = Core diameter of the bolt in mm

σ_t = Allowable tensile stress for the material of the bolts in MPa

n_b = Number of bolts (Generally, two bolts are used)

∴   Force on the bolts,

$$\boxed{F_t = \frac{\pi}{4} (d_{cb})^2 \times \sigma_t \times n_b}$$

From this expression, d_{cb} is obtained.

The nominal or major diameter (d_b) of the bolt is given by,

$$\boxed{d_b = \frac{d_{cb}}{0.84}}$$

(d) Thickness of the big end cap :

The thickness of the big end cap (t_c) may be determined as below :

Maximum bending moment acting on the cap will be taken as

$$M_c = \frac{F_I \times x}{6}$$

where, x = Distance between the bolt centres

= Diameter of crankpin or big end bearing

(d_c) + 2 × Thickness of bearing linear (3 mm) + Clearance (3 mm)

Let, b_c = Width of the cap in mm. It is equal to the length of the crank pin or big end bearing (l_c)

and σ_b = Allowable bending stress for the material of the cap in MPa

Section modulus for the cap,

$$Z_c = \frac{b_c\,(t_c)^2}{6}$$

∴   Bending stress, $\sigma_b = \dfrac{M_c}{Z_c} = \dfrac{F_t \times x}{6} \times \dfrac{6}{b_c\,(t_c)^2} = \dfrac{F_I \times x}{b_c\,(t_c)^2}$

From this expression, the value of t_c is obtained.

SOLVED PROBLEMS

Problem 4.17 : *Design a big end bolts of connecting rod with following data :*

Maximum inertia force on the connecting rod is 3000 N, at 4500 rpm

Allowable stress for bolt = 65 N/mm².   **(S-15)**

Solution : Given : F_I = 3000 N and F_t = 65 N/mm²

The bolts are under tension due to load,

$$F_I = \frac{\pi}{4}\, d_c^2 \times f_t \times 2$$

∴ $3000 = \dfrac{\pi}{4}\,(d_c)^2 \times 65 \times 2$

∴ d_c = 5.4 mm

Now,   Diameter of bolt = $\dfrac{d_c}{0.84}$

∴ $d = \dfrac{5.4}{0.84} =$ **6.45 mm**   say **7 mm**

Problem 4.18 : *Design the connecting rod cross-section with the following data of petrol engine.*

Maximum pressure inside the cylinder = 4.5 N/mm², Piston diameter = 70 mm, Stroke length = 80 mm

Effective length of connecting rod = 140 mm

Maximum allowable stress in the connecting rod in clop pin is 100 N/mm².

Take Rankine constant for steel $\dfrac{1}{1600}$.   **(S-18)**

Solution : Let thickness of the flange and web of the section = t

Width of section, B = 4t and Depth or height of the section, H = 5t

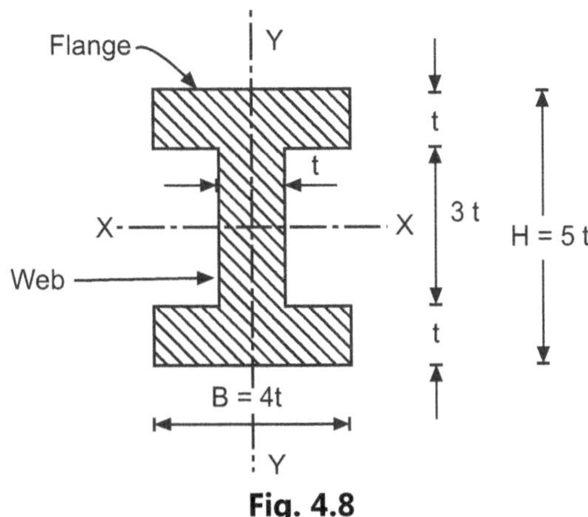

Fig. 4.8

From figure, Area of section,

$$A = 2(4t \times t) + 3t \times t$$

$$\therefore \quad A = 11t^2$$

Given Data : p_{max} = 4.5 N/mm^2

D = 70 mm

l = 80 mm

L = 140 mm

$K = \sqrt{3.18\,t^2}$

σ_{cu} = 330 N/mm^2

σ_c = 100 N/mm^2

$A = 11t^2$ where t = thickness of rod,

α = Rankine constant = $\dfrac{1}{1600}$.

$$W = \text{Maximum gas load} = p_{max} \times \frac{\pi}{4} D^2$$

$$= 4.5 \times \frac{\pi}{4} 70^2 = 17.32 \times 10^3 \text{ N}$$

Assuming I-section, $\quad W = \dfrac{\sigma_c \times A}{1 + \alpha \left[\dfrac{L^2}{K_{xx}^2}\right]}$

$$\therefore \quad t = \textbf{4.08 mm} \cong \textbf{4.5 mm}$$

Other dimensions of I-section at middle, small end and big end.

(a) At the middle or centre dimension :

(i) Depth or height of section :

$$H = 5t = 5 \times 4.5$$

$$\therefore \quad H = \textbf{22.5 mm}$$

(ii) Width of cross-section B :

$$B = 4t = 4 \times 4.5 = \textbf{18 mm}$$

(b) Dimension at small end :

(i) Depth or height of section :

$$H_1 = 0.82\,H = 0.82 \times 22.5 = \textbf{18.45 mm}$$

(ii) Width of cross-section B :

$$B = B_1 = \textbf{18 mm}$$

(c) Dimension at big end :

(i) Depth or height of section :

$$H_2 = 1.18\,H = \textbf{26.55 mm}$$

(ii) Width of cross-section B :

$$B_2 = B = \textbf{18 mm}$$

Problem 4.19 : *Determine dimensions of the cross-section of a connecting rod of an I.C. engine for following data : Cylinder bore = 100 mm, Length of connecting rod = 350 mm, Maximum gas pressure = 4 MPa, F.O.S. = 6, Rankine constant = $\dfrac{1}{7500}$. Also find variations in height of cross-section and draw a neat proportionate sketch of connecting rod.*

(W-15)

Solution : Given :　　　　　　　D = 100 mm,　F.O.S. = 6

Maximum gas pressure,　　　　p_{max} = 4 N/mm^2

Length of connecting rod,　　　L = 350 mm

1.　Force acting on the connecting rod :

$$P_c = \left(\frac{\pi D^2}{4}\right) P_{max} = \left[\frac{\pi (100)^2}{4}\right] (4) = 31415.93 \text{ N}$$

2.　Critical buckling load :

$$P_{cr} = P_c(fs) = 31415.96 \ (6) = 188495.58 \text{ N}$$

3.　Calculation of thickness, t :

Substituting,　A = 11 t^2,　K_{xx} = 1.78 t,　$\alpha = \dfrac{1}{7500}$,　σ_c = 330 N/mm^2

$$P_{cr} = \frac{\sigma_c A}{1 + \alpha \left(\dfrac{L}{K_{xx}}\right)^2}$$

$$\therefore \quad 188495.58 = \frac{(330)\,(11\ t^2)}{1 + \dfrac{1}{7500}\left(\dfrac{350}{1.78\ t}\right)^2}$$

$$\therefore \quad \frac{188495.58}{(330)\,(11)} = \frac{t^2}{1 + \dfrac{5.16}{t^2}}$$

$$\therefore \quad 51.93 = \frac{t^4}{t^2 + 5.16}$$

$$\therefore \quad t^4 - 51.93\ t^2 - 267.96 = 0$$

The above expression is a quadratic equation in (t^2).

Solving the equation, we get 　$t^2 = \dfrac{51.93 \pm \sqrt{(51.93)+2 + 4(267396)}}{2} = \dfrac{51.93 \pm 61.39}{2} = 56.66$

$$\therefore \quad t = 7.3 = 8 \text{ mm}$$

4.　Dimensions of cross-section :

Thickness of web,　　　　　　t = 8 mm

Thickness of flanges,　　　　　t = 8 mm

　　　　　　　　　　　　　B = 4t = 4(8) = 32 mm

　　　　　　　　　　　　　H = 5t = 5(8) = 40 mm

The width (B = 32 mm) is kept constant throughout the length of connecting rod.

5.　Variation of height :

At the middle section,　　　H = 5t = 5(8) = 40 mm

At the small end,　　　　　H_1 = 0.85 H = 0.85 (40) mm = 34 mm

At the big end, $H_2 = 1.2\,H = 1.2\,(40) = 48$ mm

Dimension (B/H) of section at big end = 32 mm × 48 mm

Dimensions (B/H) of section at middle = 32 mm × 48 mm

Dimensions (B/H) of section at small end = 32 mm × 48 mm

For Fig. : Refer Fig. 4.6.

Problem 4.20 : *Design connecting rod cross-section with following data :*

1. *Maximum pressure inside cylinder = 6.5 N/mm^2.*   2. *Piston diameter = 100 mm.*

3. *Stroke length = 110 mm.*   4. *Effective length of connecting rod = 220 mm.*

5. *Maximum allowable stress in crippling = 120 MPa.*   6. *Rankine constant = $\dfrac{1}{6000}$.*

Also calculate height of cross-section at both the ends.   **(S-16)**

Solution : Given : $p_{max} = 6.5$ N/mm^2, D = 100 mm, I = 110 mm, L = 220 mm,

$$\sigma_c = 120 \text{ N/mm}^2$$

$$A = 11\,t^2 \text{ where } t = \text{thickness of rod}$$

$$\alpha = \text{Rankine constant} = \frac{1}{6000}$$

$$K_{xx} = \sqrt{3.18\,t^2}$$

Maximum gas load, $$W = p_{max} \times \frac{\pi}{4} D^2 = 6.5 \times \frac{\pi}{4}\,100^2 = 51.05 \times 10^3 \text{ N}$$

Assuming I-section, $$W = \frac{\sigma_c \times A}{1 + \alpha\left[\dfrac{L^2}{K_{xx}^2}\right]}$$

$\therefore$ $$51.05 \times 10^3 \text{ N} = \frac{120 \times 11\,t^2}{1 + \dfrac{1}{6000}\left[\dfrac{220^2}{3.18\,t^2}\right]}$$

$\therefore$ $$t^4 - 38.674\,t^2 - 98.07 = 0$$

This is quadratic equation in t^2 $$t^2 = 41.19$$

$\therefore$ $$t = \mathbf{6.418 \text{ mm}} \cong \mathbf{6.5 \text{ mm}}$$

Other dimensions of I-section at middle, small end and big end :

1. Dimensions at the middle or center :

(i)   Depth or height of section, $H = 5t = 5 \times 6.5$

$$H = \mathbf{32.5 \text{ mm}}$$

(ii) Width of cross-section, $B = 4t = 4 \times 6.5 = \mathbf{26 \text{ mm}}$

2. Dimensions at small end :

(i)   Depth or height of section, $H_1 = 0.82\,H = 0.82 \times 32.5 = \mathbf{26.65 \text{ mm}}$

(ii) Width of cross-section, $B = B_1 = \mathbf{26 \text{ mm}}$

3. Dimensions at big end :

(i)   Depth or height of section, $H_2 = 1.182\,H = 1.182 \times 32.5 = \mathbf{38.35 \text{ mm}}$

(ii) Width of cross-section, $B = B_2 = \mathbf{26 \text{ mm}}$

4.4 DESIGN OF ROCKER ARM (S-16, W-17)

Q.1. *Describe the procedure to design fulcrum pin of rocker arm.* (S-14, W-14)

Q.2. *Write design procedure of a rocker arm lever for cross-section only.* (S-15)

Design Procedure of a Rocker Arm for Operating Exhaust Valve :

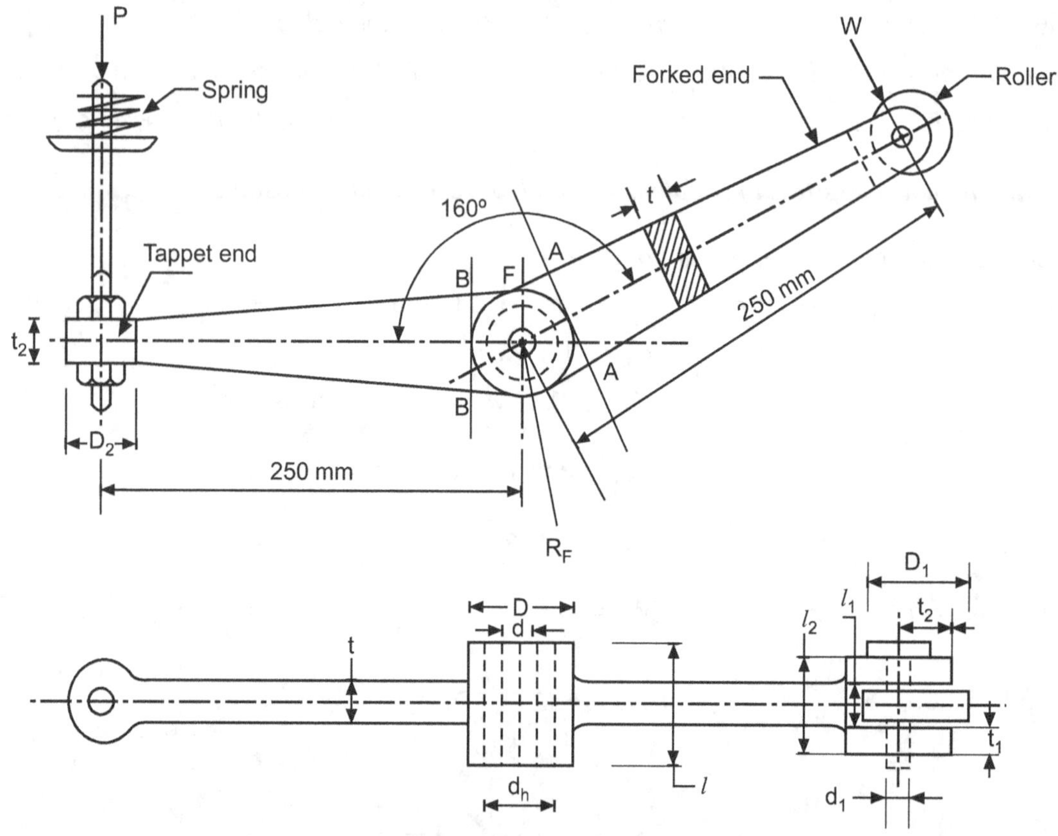

Fig. 4.10

In designing a rocker arm the following procedure may be followed :

1. Rocker arm is usually I-section. It is subjected to bending moment. To find bending moment, it is assumed that the arm of the lever extends from point of application of load to centre of pivot.

2. The ratio of length to the diameter of the fulcrum pin and roller pin is taken as 1.25. The permissible bearing pressure on this pin is taken from 3.5 to 6 N/mm^2.

3. The outside diameter of the boss of fulcrum is usually taken twice the diameter of the pin at fulcrum. The boss is provided with a 3 mm thick phosphor bronze bush to take up the wear.

4. One end of rocker arm has a forked end to receive roller.

5. The outside diameter of the eye at the forked end is also taken as twice the diameter of pin. The diameter of roller is slightly larger (at least 3 mm more) than the diameter of eye at the forked end. The radial thickness of each eye of the forked end is taken half the diameter of pin. Some clearance about 1.5 mm must be provided between the roller and the eye at the forked end so that roller can move freely. The pin should, therefore be checked for bending.

6. The other end of rocker arm (i.e. tappet end) is made circular to receive the tappet which is a stud with a lock nut. The outside diameter of the circular arm is taken as twice the diameter of the stud. The depth of section is also taken twice the diameter of stud.

Design Procedure of Fulcrum Pin of Rocker Arm :

Step 1 : Calculate reaction at the fulcrum pin :

$$R_F = \sqrt{W^2 + P^2 - 2W \times P \times \cos \theta}$$

Step 2 : Design of fulcrum pin :

(a)   Let, d = Diameter of the fulcrum pin

        and l = Length of the fulcrum pin = 1.25 d

Considering the bearing of the fulcrum pin.

Load on the fulcrum pin (R_F) :

$$\text{Bearing pressure} = \frac{\text{Load}}{\text{Bearing area}} = \frac{R_F}{l \times d}$$

∴ $$\boxed{p_b = \frac{R_F}{1.25\, d \times d}}$$ … (4.11)

From equation (4.11), l and d can be determined.

(b)   Checking shear stress induced in the fulcrum pin.

As the pin is in double shear,

$$\tau = \frac{R_F}{2 \times \left(\dfrac{\pi}{4} \cdot d^2\right)}$$

External diameter of the boss, $D = 2d$

Internal diameter of the hole in the lever

$$d_h = d + 2 \times 3$$

Check the induced bending stress for the section of the boss at the fulcrum.

Bending moment at this section = $W \times L$

Section modulus, $$Z = \frac{1}{12} \times l \times (D^3 - d_h^3)\, D/2$$

Induced bending stress, $$\boxed{\sigma_b = \frac{M}{Z}}$$

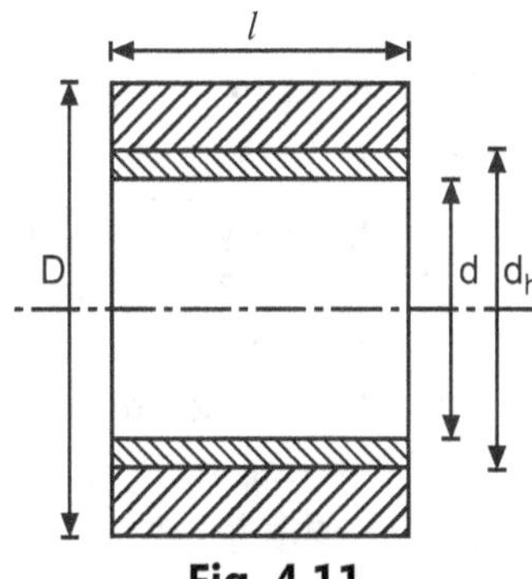

Fig. 4.11

SOLVED PROBLEM

Problem 4.21 : *Two arms of rocker arm are equal and make included angle of 160°. It is used to operate exhaust valve, for which maximum force required is 5 kN. Determine length of fulcrum pin and induced shear stress in pin material, if allowable bearing pressure in pin material is 7 N/mm².*   **(W-15)**

Solution : Given : $\theta = 160°$

                                        $P = 5 \text{ kN}$

                                        $p_b = 7 \text{ N/mm}^2$

Two arms of rocker arm are equal.

∴ $L_1 = L_2$

and $P = W = 5 \text{ kN}$

1.   Reaction at fulcrum pin :

$$R_f = \sqrt{P^2 + W^2 - 2PW \cos \theta} = \sqrt{5^2 + 5^2 - 2 \times 25 \cos 160}$$

∴ $R_f = \mathbf{9.848 \text{ kN} = 9.848 \times 10^3 \text{ N}}$

2. **Diameter of fulcrum pin :**

$$R_f = d \times l \times p_b$$

$\therefore \quad 9.848 \times 10^3 = d \times 1.25\, d \times 7 \qquad (\because l = 1.25\, d)$

$\therefore \quad d = \textbf{33.54 mm} \cong \textbf{34 mm}$

3. **Shear stress induced in pin :**

$$R_f = 2 \times \frac{\pi}{4} d^2 \times \sigma_s$$

$\therefore \quad 9.848 \times 10^3 = 2 \times \frac{\pi}{4} 34^2 \times \sigma_s$

$\therefore \quad \sigma_s = \textbf{5.423 N/mm}^2$

4. **Length of fulcrum pin :** $\quad l = 1.25\, d$

$\therefore \quad l = 1.25 \times 34 = \textbf{42.5 mm} \cong \textbf{43 mm}$

4.5 FUNCTION OF VALVE SPRING, MATERIALS WITH JUSTIFICATION AND DESIGN OF VALVE SPRING

4.5.1 Function of Valve Spring in an Engine

A helical spring is used to hold closed a valve in the cylinder head of an internal combustion engine. Any spring that closes a valve after it has been opened mechanically or by flow pressure.

The function of an internal combustion engine valve spring is to provide sufficient force throughout the engine cycle to maintain the tappet in contact with the cam at all speeds within the engine speed range. However, it must not exert such force that will produce unacceptable contact stresses between tappet and cam.

The performance requirements of a valve spring are that it must not fail by fracture or by load relaxation, lose so much of its controlling force that it fails to maintain control between tappet and cam.

4.5.2 Material for Valve Spring

(i) High carbon spring steel,
(ii) Alloy spring steel,
(iii) Stainless steel,
(iv) Phosphor bronze and copper base alloys.

Above material is used for valve spring because of their better compressibility, high strength, high stiffness.

4.5.3 Design Procedure

A valve spring is a compression type spring that is kept around a valve stem and retained by a part called a retainer. The length of valve movement, called lift distance, determines how much the spring will be compressed. An external force on the valve spring opens the valves and the springs get compressed.

Essential equations for compression spring Design calculation

The following equations will be used for designing or selecting a valve compression spring.

(i) Relation between the applied external force and generated maximum shear stress in the wire :

$$f_{max} = \frac{W \times (8FD)}{\pi \times d^3} \qquad \ldots (4.12)$$

where, $\qquad f_{max}$ = Maximum shear stress generated in wire

$\qquad w$ = Wahl's correction factor

$\qquad F$ = Applied external force

$$D = \text{Nominal diameter of the spring coil}$$
$$D = \text{Wire diameter of the spring}$$

Further, W can be calculated as,

$$W = \frac{(4C - 1)}{(4C - 4) + \left(\dfrac{0.615}{C}\right)} \qquad \ldots (4.13)$$

Where, $C = \text{Spring index}\left(\text{i.e. } C = \dfrac{d}{D}\right)$

(ii) Relation between number of active coils and spring constant

$$N_a = \frac{(G \times d^4)}{(8 \times D^3 \times K)} \qquad \ldots (4.14)$$

Where,

$$N_a = \text{Required number of active coils}$$
$$G = \text{Shear modulus of the spring material}$$
$$d = \text{Spring wire diameter}$$
$$D = \text{Nominal Diameter of the spring}$$
$$K = \text{Spring constant}$$

Steps of Valve Compression Spring Design :

(1) Either applied external force (F) or maximum generated shear stress in spring wire (f_{max})

(2) Value of compression (δ) under externally applied force.

(3) Any two values among spring wire diameter (d), nominal spring diameter (D) and spring index (C).

(4) Shear modulus of the spring material (G).

- Use equation (2) for calculating Wahl correction factor (w)
- Use equation (1) and find out unknown of F and f_{max}
- Calculate the valve spring stiffness (K) using the equation

$$F = K \times \delta$$

- Use equation 3 for calculating required number of spring coil (Na)

SOLVED PROBLEM

Problem 4.22 : *A helical spring is made from a wire of 6 mm diameter and has outside diameter of 75 mm. If the permissible shear stress is 350 MPa and modulus of rigidity 84 kN/mm², find the axial load which the spring can carry and the deflection per active turn.*

Solution : Given : d = 6 mm, D_o = 75 mm, τ = 350 MPa = 350 N/mm²,

$$G = 84 \text{ kN/mm}^2 = 84 \times 10^3 \text{ N/mm}^2$$

We know that, mean diameter of the spring,

$$D = D_o - d = 75 - 6 = 69 \text{ mm}$$

$\therefore$ Spring index, $C = \dfrac{D}{d} = \dfrac{69}{6} = 11.5$

Let, $W = \text{Axial load and}$

$$\delta/n = \text{Deflection per active turn}$$

(1) Neglecting the effect of curvature : We know that the shear stress factor,

$$K_s = 1 + \frac{1}{2C} = 1 + \frac{1}{2 \times 11.5} = 1.043$$

and maximum shear stress induced in the wire (τ),

$$350 = K_s \times \frac{8 \cdot W \cdot D}{\pi \times d^3} = 1.043 \times \frac{8W \times 69}{\pi \times 6^3} = 0.848 \, W$$

$$\therefore \quad W = \frac{350}{0.848} = \textbf{412.7 N}$$

We know that deflection of the spring,

$$\delta = \frac{8WD^3 n}{G.d^4}$$

$\therefore$ Deflection per active turn,

$$\frac{\delta}{n} = \frac{8W \cdot D^3}{Gd^4} = \frac{8 \times 412.7 \,(69)^3}{84 \times 10^3 \times 6^4} = \textbf{9.96 mm}$$

(2) Considering the effect of curvature : We know that, Wahl's stress factor,

$$K = \frac{4C - 1}{4C - 4} + \frac{0.615}{C} = \frac{4 \times 11.5 - 1}{4 \times 11.5 - 4} + \frac{0.615}{11.5} = 1.123$$

We also know that, the maximum shear stress induced in the wire (τ),

$$350 = K \times \frac{8W \cdot C}{\pi d^2} = 1.123 \times \frac{8 \times W \times 11.5}{\pi \times 6^2} = 0.913 \, W$$

$$\therefore \quad W = \frac{350}{0.913} = \textbf{383.4 N}$$

and deflection of the spring, $\delta = \dfrac{8W \cdot D^3 \cdot n}{Gd^4}$

$\therefore$ Deflection per active turn,

$$\frac{\delta}{n} = \frac{8 \cdot W \cdot D^3}{Gd^4} = \frac{8 \times 383.4 \,(69)^3}{84 \times 10^3 \times 6^4} = \textbf{9.26 mm}$$

4.6 PUSH ROD

4.6.1 Function of Push Rod

Push rod connects the camshaft and the rocker arm to convert rotary motion of the camshaft into pulse motion of the rocker arm. Basically, it is served as a purpose of valve operation for IC engine. It acts as a linkage to execute the valve opening and closing for IC engine.

4.6.2 Pushed Material

Push rods are typically of 1018 steel tubing. Other materials are 4130 or 4140. Chromoly tubing are much stronger and premature wear.

4.6.3 Design Procedure of Push Rod

The push rods are used in overhead valve and side valve engines. Since these are designed as long columns, therefore Euler's formula should be used. The push rods may be treated as pin end columns because they use spherical seated bearings.

Let

W = Load acting on the push rod,

D = Diameter of the push rod,

d = Diameter of the hole through the push rod,

I = Moment of inertia of the push rod,

$\quad = \dfrac{\pi}{64} \times D^4$, for solid rod $= \dfrac{\pi}{64}(D^4 - d^4)$, for hollow rod

l = Length of the push rod, and

E = Young's modulus for the material of push rod.

If m is the factor of safety for the long columns, then the critical or crippling load on the rod is given by

$$\boxed{W_{cr} = m \times W}$$

Now, using Euler's formula, $W_{cr} = \dfrac{\pi^2 EI}{L^2}$, the diameter of the push rod (D) can be obtained.

- Generally, the diameter of the hole through the push rod is 0.8 times the diameter of push rod i.e. **d = 0.8 D**
- Since push rods are treated as pin end columns, therefore the equivalent length of the rod (L) is equal to the actual length of the rod (l).

SOLVED PROBLEMS

Problem 4.23 : *The maximum load on a petrol engine push rod 300 mm long is 1400 N. It is hollow having the outer diameter 1.25 times the inner diameter. Spherical seated bearings are used for the push rod. The modulus of elasticity for the material of the push rod is 210 kN/mm². Find a suitable size for the push rod, taking a factor of safety of 2.5.*

Given : l = 300 mm, W = 1400 N, D_o = 1.25 D_i, E = 210 kN/mm² = 210×10^3 N/mm².

F.O.S. = 2.5.

To find : D_o and D_i.

Solution : Since, the push rods are treated as pin end columns, therefore the equivalent length of the rod (L_e) is equal to the actual length of the rod (l).

$$\therefore \qquad L_e = l = 300 \text{ mm}$$

Moment of inertia for a hollow section is,

$$I = \frac{\pi}{64} \times [D_o^4 - D_i^4] = \frac{\pi}{64} [(1.25\, D_i)^4 - D_i^4] = 0.070\, D_i^4$$

Crippling load on the push rod,

$$W = W \times F.O.S. = 1400 \times 2.5$$

$$\therefore \qquad W_{cr} = 3500 \text{ N}$$

Now, according to Euler's formula Crippling load is given by,

$$W_{cr} = \frac{\pi^2 E l}{(L_e)^2} = \frac{\pi^2 \times 210 \times 10^3 \times 0.070\, D_i^4}{(300)^2} = 1.612\, D_i^4$$

$$\therefore \qquad 3500 = 1.612\, D_i^4$$

$$\therefore \qquad D_i = 6.82 \text{ mm} \qquad\qquad \therefore \ D_i \cong \textbf{8 mm}$$

Outer diameter of push rod,

$$D_o = 1.25\, D_i = 1.25 \times 8 = 10 \text{ mm} \qquad \therefore\ D_o = \textbf{10 mm}$$

Problem 4.24 : *Determine the diameter of the push rod made of mild steel of an I.C. engine if the maximum force exerted by the push rod is 1500 N. The length of the push rod is 0.5 m. Take the factor of safety as 2.5 and the end fixity coefficient as 2.*

Given : W = 1500 N, l = 0.5 m = 500 mm, F.O.S. = 2.5, C = 2.

To find : D = ?

Solution : Since the push rods are treated as pin end columns, therefore the equivalent length of the rod (L_e) is equal to the actual length of the rod (l).

$$\therefore \qquad L_e = l = 500 \text{ mm}$$

Moment of inertia for a solid section is,

$$I = \frac{\pi}{64} \times D^4 = 0.049\, D^4$$

Crippling load on the push rod,

$$W_{cr} = W \times \text{F.O.S.} = 1500 \times 2.5 = 3750 \text{ N}$$

Now, according to Euler's formula Crippling Load is given by,

$$W_{cr} = \frac{C\pi^2 EI}{(L_e)^2} \qquad \text{(where, C = end fixity coefficient)}$$

$$= \frac{2 \times \pi^2 \times 210 \times 10^3 \times 0.049 \, D^4}{(500)^2} \qquad \text{(Assuming, E} = 210 \times 10^3 \text{ N/mm}^2\text{)}$$

$$\therefore \qquad W_{cr} = 0.8124 \, D^4$$

$$\therefore \qquad 3750 = 0.8124 \, D^4$$

$$\therefore \qquad D = 8.24 \text{ mm} \qquad \therefore \; D \cong \textbf{10 mm}$$

Problem 4.25 : The maximum load on a petrol engine push rod 400 mm long is 1800 N. It is hollow having the outer diameter 1.5 times the inner diameter. Spherical seated bearings are used for the push rod. The modulus of elasticity for the material of the push rod is 200 kN/mm^2. Find a suitable size for the push rod. Take a factor of safety is 2.5 and the end fixity coefficient as 2.

Given : $l = 400$ mm, $W = 1800$ N, $D_o = 1.5 \, D_i$, $E = 200$ kN/mm$^2 = 200 \times 10^3$ N/mm^2,

F.O.S. = 2.5,   C = 2

To find : D_o and D_i

Solution : Since, the push rods are treated as pin end columns, therefore the equivalent length of the rod (L_e) is equal to the actual length of the rod (l).

$$\therefore \qquad L_e = l = 400 \text{ mm}$$

Moment of inertia for a hollow section is,

$$I = \frac{\pi}{64} \times [D_o^4 - D_i^4] = \frac{\pi}{64} \times [(1.5 \, D_i)^4 - D_i^4] \qquad \therefore \; I = 0.20 \, D_i^4$$

Crippling load on the push rod,

$$W_{cr} = W \times \text{F.O.S.} = 1800 \times 2.5 = 4500 \text{ N}$$

Now, according to Euler's formula crippling load is given by,

$$W_{cr} = \frac{C\pi^2 EI}{(L_e)^2} = \frac{2 \times \pi^2 \times 200 \times 10^3 \times 0.20 \, D_i^4}{(400)^2} = 4.934 \, D_i^4$$

$$\therefore \qquad 4500 = 4.934 \, D_i^4$$

$$\therefore \qquad D_i = 5.49 \text{ mm} \qquad\qquad \therefore \; D_i \cong \textbf{6 mm}$$

Outer diameter of push rod,

$$D_o = 1.5 \, D_i = 1.5 \times 6 = 9 \text{ mm} \qquad \therefore \; D_o = \textbf{9 mm}$$

Problem 4.26 : Determine the diameter of the push rod made of mild steel of an I.C. engine if the maximum force exerted by the push rod is 1800 N. The length of the push rod is 0.45 m. Take the factor of safety as 3 and $E = 210 \times 10^3$ N/mm^2.

Given : $W = 1800$ N, $l = 0.45$ m $= 450$ mm,   F.O.S. = 3

To find : $D = ?$

Solution : Since, the push rods are treated as pin and columns, therefore the equivalent length of the rod (L_e) is equal to the actual length of the rod (l).

$$\therefore \qquad L_e = l = 450 \text{ mm}$$

Moment of inertia for a solid section is,

$$I = \frac{\pi}{64} \times D^4 = 0.049 \, D^4$$

Crippling load on the push rod,

$$W_{cr} = W \times \text{F.O.S.} = 1800 \times 3 = 5400 \text{ N}$$

Now, according to Euler's formula Crippling load is given by,

$$W_{cr} = \frac{\pi^2 \, E \, I}{(L_e)^2} = \frac{\pi^2 \times 210 \times 10^3 \times 0.049 \, D^4}{(450)^2} = 0.50 \, D^4$$

$$\therefore \quad 5400 = 0.50 \, D^4$$

$$\therefore \quad D = 10.19 \text{ mm} \cong \textbf{12 mm}$$

Practice Questions

1. Explain design of piston crown for bending strength.
2. Describe the procedure to design connecting rod cross-section (I-section).
3. Explain indicated power and brake power of an engine cylinder.
4. Explain design of piston pin on the basis of bearing pressure and shear strength.

Problems for Practice

Problems on Cylinder and Cylinder Head :

1. A four stroke internal combustion engine has the following specifications.
   Brake power = 7.5 kW, Speed = 1000 r.p.m., Indicated mean effective pressure = 0.35 N/mm^2
   Maximum gas pressure = 3.5 N/mm^2, Mechanical efficiency = 80%
2. Determine :
   (1) The dimensions of the cylinder, if the length of stroke is 1.4 times the bore of the cylinder.
   (2) Wall thickness of the cylinder, if the hoopstron or permissible circumferential stress is 35 MPa.
   (3) Thickness of the cylinder head and size of studs when the permissible stresses for the cylinder head and stud material are 45 MPa and 65 MPa respectively.

Problem on Piston Head or Crown and Piston Pin :

3. Design a piston for a four stroke diesel engine consuming 0.3 kg of the fuel per kW of power per hour and produces a brake mean effective pressure of the 0.7 N/mm^2. The maximum gas pressure inside the cylinder is 5N/mm^2 at a speed of 3200 rpm. The cylinder diameter required is to be 280 mm with stroke 1.4 times the diameter. The piston may have 4 compression rings and an oil ring.
   C_v = Calorific value of fuel = 4.6 × 10^3 kJ/kg
   Temperature at the piston centre = 700°K
   Temperature at the piston edge = 470°K
   Heat conductivity factor (k) = 46.6 W/m/°K
   Heat conducted through top = 5 % of heat produced
   Permissible tensile strength for the material of the piston = 27 N/mm^2
   Permissible tensile stress in rings = 80 N/mm^2
   Permissible pressure on piston barrel = 0.4 N/mm^2
   Permissible pressure on piston pin = 15 N/mm^2
   Permissible stress in piston pin = 80 N/mm^2

Problem on Connecting Rod :

4. The following data is given for designing a connecting rod used for high speed, 4 stroke I.C. engine.
   Diameter of piston = 82 mm
   Mass of reciprocating parts = 1.5 kg
   Length of connecting rod = 310 mm
   Stroke = 125 mm
   R.P.M. = 2200
   Compression ratio = 6.8 : 1
   Explosion pressure = 3.5 N/mm^2

Summer 2016

1. Define : (1) Indicated power, (2) Brake power,

    (3) Frictional power and state relation between them.   **(4 M)**

Ans. Refer Article 4.1.2.

2. State any four requirements of piston material.   **(4 M)**

Ans. Refer Article 4.2.

3. A four stroke diesel engine has following specifications :   **(4 M)**

    1. B.P. – 5 kW at 1200 rpm.

    2. Indicated mean effective pressure 0.35 N/mm^2.

    3. Mechanical efficiency 80%.

    Determine :

    1. Bore and length of cylinder.

    2. Thickness of cylinder head.

    3. Size of studs for cylinder head if allowable tensile strength for stud material is 65 MPa.   **(8 M)**

Ans. Refer Solved Problem 4.2.

4. Design piston pin for following data : piston diameter - 70 mm, maximum gas pressure inside cylinder - 4.5 N/mm^2. Allowable stresses in bending, shear and bearing are 100 MPa, 70 MPa and 25 MPa respectively.   **(8 M)**

Ans. Refer Solved Problem 4.16.

5. Design connecting rod cross-section with following data :

    1. Maximum pressure inside cylinder 6.5 N/mm^2.

    2. Piston diameter = 100 mm.

    3. Stroke length = 110 mm.

    4. Effective length of connecting rod = 220 mm.

    5. Maximum allowable stress in crippling = 120 MPa.

    6. Rankine's constant $= \dfrac{1}{6000}$.

    Also calculate height of cross-section at both the ends.   **(8 M)**

Ans. Refer Solved Problem 4.14.

6. State the design procedure for piston rings and skirt length.   **(8 M)**

Ans. Refer Article 4.2.2 and 4.2.3.

Winter 2016

1. Describe the stepwise procedure for designing the piston crown of an engine for bending strength and thermal consideration.   **(4 M)**

Ans. Refer Article 4.2.1.

2. Explain design procedure of a connecting rod.   **(6 M)**

Ans. Refer Article 4.3.1.

3. Describe in detail the design procedure used to design the piston rings and piston skirts.   **(8 M)**

Ans. Refer Articles 4.2.2 and 4.2.3.

4. Describe in detail design procedure to design :

    (i) Thickness of cylinder head,   (ii) Cylinder head bolts or studs.   **(8 M)**

Ans. Refer Article 4.1.

Summer 2017

1. State the material used for following components with proper justification :
   (i) Piston,   (ii) Crank shaft.   **(4 M)**
Ans. Refer Article 4.2.

2. With neat sketch, show that thrust side and non-thrust side of I.C. engine piston.   **(4 M)**
Ans. Refer Fig. 4.3.

3. Determine the thickness of plain cylinder head for 300 mm cylinder diameter. The maximum gas pressure is 3.2 N/mm^2. Take allowable tensile stress for cylinder cover is 42 N/mm^2 and constant is 0.1.   **(4 M)**
Ans. Refer Solved Problem 4.3.

4. State any four design considerations for design of piston.   **(4 M)**
Ans. Refer Article 4.2.

5. Write design procedure for connecting rod.   **(8 M)**
Ans. Refer Article 4.3.1.

6. A four stroke diesel engine has the following specifications :
   Brake power = 5 kW, Speed = 1200 rpm, Indicated mean effective pressure = 0.35 N/mm^2,
   Mechanical efficiency = 80%.
   Determine : Bore and length of cylinder, Thickness of cylinder head.
   Assume length of stroke, l = 1.5 D, or l = 1.08 D, Constant (C) = 0.1
   Tensile stress for cylinder cover = 52 N/mm^2.   **(8 M)**
Ans. Refer Solved Problem 4.2.

7. Design the piston pin with following data :
   (i)   Maximum pressure on piston= 4 N/mm^2.
   (ii) Diameter of piston = 70 mm.
   (iii) Allowable stresses due to bearing, bending and shear are 30 N/mm^2, 80 N/mm^2 and 60 N/mm^2 respectively.   **(8 M)**
Ans. Refer Solved Problem 4.15.

Winter 2017

1. Determine the bore and length of cylinder of 4-stroke diesel engine for following specification - Brake power - 5 kW, Speed - 1200 r.p.m., P_m - 0.35 N/mm^2, Mechanical efficiency - 80%, L/D = 1.08.   **(4 M)**
Ans. Refer Solved Problem 4.4.

2. Explain the design procedure of Rocker arm for operating exhaust valve.   **(6 M)**
Ans. Refer Article 4.4.

3. Determine the thickness of plain cylinder head for 0.3 m cylinder diameter. The maximum gas pressure is 3.2 N/mm^2. Take C = 0.1, σ_1 = Tensile stress = 42 N/mm^2.   **(4 M)**
Ans. Refer Solved Problem 4.3.

4. Write the design procedure for designing of piston head or crown by strength consideration.   **(4 M)**
Ans. Refer Article 4.2.1.

5. Write the design procedure for connecting rod.   **(6 M)**
Ans. Refer Article 4.3.1.

6. Design a piston pin with following data : Maximum gas pressure = 4 N/mm^2, Diameter of piston = 70 mm, Allowable stresses due to bearing, bending and shear are 30 N/mm^2, 80 N/mm^2, 60 N/mm^2 respectively.   **(8 M)**
Ans. Refer Solved Problem 4.15.

7. Design piston ring for following : Number of ring = 5, Wall pressure (P_w) = 0.035 N/mm^2, Bending stress for ring = 85 N/mm^2, Diameter of cylinder bore = 240 mm.   **(4 M)**
Ans. Refer Solved Problem 4.11.

8. Design the skirt length of the piston with the following data of petrol engine. Maximum pressure inside the cylinder = 6.5 N/mm^2. Piston diameter = 100 mm, Side thrust is limited to 10% of maximum load on the piston. Allowable bearing pressure = 0.3 N/mm^2. **(4 M)**

Ans. Refer Solved Problem 4.13.

Summer 2018

1. Define : (i) Indicated power, (ii) Brake power, (iii) Frictional power and state relation between them. **(4 M)**

Ans. Refer Article 4.1.2.

2. A four stroke diesel engine has the following specifications :
   Brake power = 5 kW, Speed = 1200 rpm, Indicated mean effective pressure = 0.35 N/mm^2, Mechanical efficiency = 80%
   Determine : (i) Bore and length cylinder, (ii) Thickness of cylinder head. **(6 M)**

Ans. Refer Solved Problem 4.2.

3. Design piston pin with following data :
   Maximum gas pressure = 4 N/mm^2, Diameter of piston = 70 mm.
   Allowable stresses due to bearing, bending and shear are given 30 N/mm^2, 80 N/mm^2, 60 N/mm^2 respectively. **(8 M)**

Ans. Refer Solved Problem 4.15.

4. Explain the design procedure used to design the piston rings and piston skirts. **(8 M)**

Ans. Refer Articles 4.2.2 and 4.2.3.

5. Design the connecting rod cross-section with the following data of petrol engine.
   Maximum pressure inside the cylinder = 4.5 N/mm^2, Piston diameter = 70 mm,
   Stroke length = 80 mm, Effective length of connecting rod = 140 mm
   Maximum allowable stress in the connecting rod in crippling is 100 N/mm^2.

   Take Rankine constant for steel $\frac{1}{600}$. **(8 M)**

Ans. Refer Problem 4.18.

Winter 2018

1. Draw thrust and non-thrust sides of I.C. engine piston. **(4 M)**

Ans. Refer Fig. 4.3.

2. Design the piston crown thickness from the following data : Diameter of piston = 90 mm, Maximum pressure on the piston = 4.6 N/mm^2 and Allowable bending stress = 45 N/mm^2. **(4 M)**

Ans. Refer Solved Problem 4.8.

3. Define indicated power and brake power of an engine cylinder. **(4 M)**

Ans. Refer Article 4.1.2.

4. Explain the design procedure of connecting rod. **(8 M)**

Ans. Refer Article 4.3.

5. Determine the thickness of plain cylinder head for 0.4 m cylinder diameter. The maximum gas pressure is 3.2 N/mm^2. Design the studs and cylinder cover. Take allowable tensile stress for cylinder cover and bolt equal to 42 N/mm^2 and 63 N/mm^2 respectively. **(8 M)**

Ans. Refer Solved Problem 4.6.

6. A 4-stroke diesel engine has the following specifications :
   Brake power = 6 kW, Speed = 1200 r.p.m., Indicated mean effective pressure = 0.35 N/mm^2, Mechanical efficiency = 80%.
   Determine : (a) Bore and length of cylinder, (b) Thickness of cylinder head. **(8 M)**

Ans. Refer Solved Problem 4.5.

✍ ✍ ✍

Chapter 5

DESIGN OF AXLES

Syllabus

5.1   Function of Front Axle. Material with Justification and Design of Front Axle

5.2   Function of Rear Axle, Material for Rear Axle with Justification and Design of Full Floating Rear Axle

About this Chapter

After reading this chapter, students will be able to :

- Justify the selection of material for the given axle.
- Explain stepwise design procedure for the given axle.
- Calculate dimensions of the cross-section of the given axle from the given data.
- Draw proportionate diagram of the given axle.

5.1 FUNCTION OF FRONT AXLE, MATERIAL WITH JUSTIFICATION AND DESIGN OF FRONT AXLE

5.1.1 Function of Front Axle

Front axle carries the weight of the front part of the automobile as well as facilities steering and absorbs shocks due to road surface variations.

The front axles are generally dead axles, but are live axles in small cars of compact designs and also in case of four wheel drive.

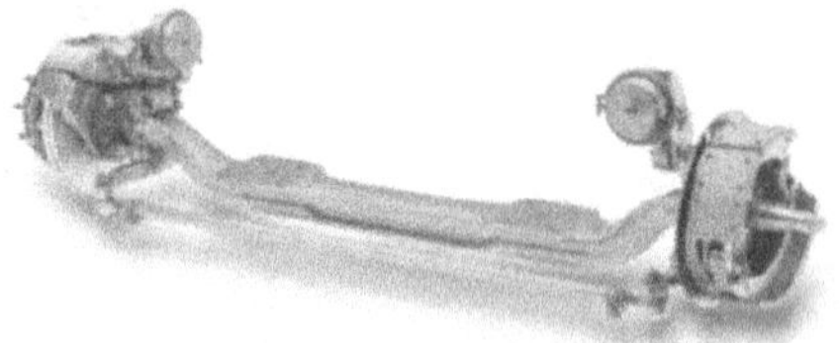

Fig. 5.1 : Front axle

5.1.2 Material

The axle beam is use is of I or H section and is manufactured from alloy forged steel for rigidity and strength.

5.1.3 Design Procedure

Front axles are subjected to both bending and shear stresses. In static condition, the axle may be considered as beam supported vertically upward at the ends i.e. at the center of the wheels and loaded vertically downward at the centers of the spring pads.

The vertical bending moment thus caused is zero at the point of support and rises linearly to a maximum at the point of loading and then remains constant.

Thus, maximum bending moment = $W \times L$ Nm

Where,   W = The load on wheel in N,

                    L = The distance between the center of wheel and the spring pad in m.

Under dynamic conditions, the vertical bending moment is increased due to road roughness. But its estimate is difficult and hence is generally accounted for through factor of safety. The front axle also experiences a horizontal bending moment because of resistant to motion.

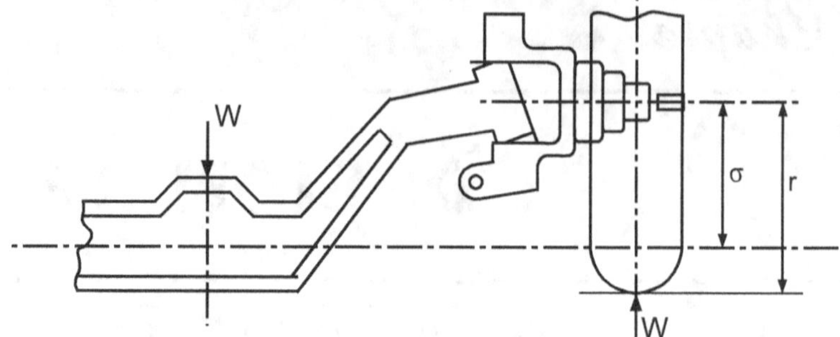

Fig. 5.2 : Loads on front axle

The magnitude of the torque = R · δ Nm.

Where,　　　　R = The resistance to motion in N,

　　　　δ = The drop from the spindle axis to the centre of the section, m.

The shear stress in the axle is due to braking torque and its magnitude as shown in Fig. 5.2 above.

Breaking torque = μ · W · r Nm

Where,　　　　r = The road wheel radius, m

　　　　μ = coefficient of friction between road and tyre

　　　　= 0.6 for dry, hard road surface

The breaking torque is lower for the section lying between the spring pads and is given by,

$$= \mu W(r - \delta)$$

In this portion, the bending moment predominates whereas at the steering head, torsion predominates. Thus, I – section is used for the portion where bending moment predominates and is gradually changed to circular, oval or rectangular section at the steering head.

For I-section, the maximum bending moment is given by the relation.

$$\boxed{\frac{M}{I} = \frac{\sigma_b}{y}}$$

Where,　　　　M = The maximum bending moment, Nm

　　　　σ_b = Allowable bending stress for the material, N/m^2

　　　　y = The maximum distance of the fiber from the neutral axis (NA)

　　　　$= \dfrac{d}{2}$ m.

　　　　I = The moment of inertia of the I-section at NA.

　　　　$I = \left(\dfrac{bd^3 - ch^3}{12}\right)$ m^4

Where,　　　　d = The overall depth of I-section, m

　　　　b = The flange width, m

　　　　t = The flange thickness, m

　　　　w = The web thickness, m

　　　　C = b – t

　　　　h = d – 2w

Generally,　　　　d = 6t

　　　　b = 4.25 t

　　　　w = 2.5 t

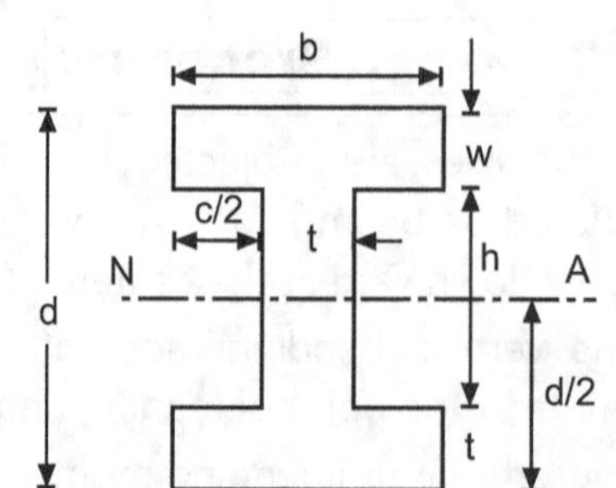

Fig. 5.3 : I-section on the axle beam

For the circular or oval section, the maximum torsion is given by the relation,

$$\boxed{\dfrac{T}{I_p} = \dfrac{f_s}{y}}$$

Where,

T = The maximum torque in the plane of section, Nm

f_s = Allowable shear stress in the material, N/m^2

y = The distance from the neutral axis to the outermost fiber of the axle.

$\quad = \dfrac{d}{2}$, m

d = The diameter for the circular section

$\quad =$ The major axis for the oval section.

I_p = Polar moment of inertia of the section

$\quad = \dfrac{\pi}{32}\, d^4$ for circular section.

$\quad = \dfrac{\pi}{32}\, d^3 b$ for oval section with minor axis, b.

SOLVED PROBLEMS

Problem 5.1 : Design a suitable I-section for the front axle assuming the following data;

Total weight of car = 13734 N,   Load taken by front axle = 6377 N,   Wheel track = 1.4 m,

Distance between the center of the spring pads = 0.7 m,

Width of flange and thickness are 0.6 and 0.2 of the overall depth of the section.

Thickness of web = 0.25 of width of flange,   Working stress = 88.29×10^6 N/m^2.

Given : W_{Total} = 13734 N,   Load on front axle = 6377 N,   Wheel track = 1.4 m,

$\quad$ b = 0.6 d,   w = 0.2 d,   t = 0.25 b, $\sigma_b = 88.29 \times 10^6$ N/m^2.

Solution : Load on each wheel of front axle is,

$$W = \frac{\text{Load taken by front axle}}{2} = \frac{6377}{2} = 3188.5 \text{ N}$$

Assuming the axle behave like a simply supported beam as shown in Fig. 5.4.

Therefore, maximum bending moment,

$$M = W \times l = 3188.5 \times 0.35 = 1115.98 \text{ N-m}$$

Assuming overall depth of I-section = 'd' m

∴ $\quad$ Width of flange, $\qquad$ b = 0.6 d

$\quad$ Thickness of flange, $\quad$ w = 0.2 d

$\quad$ Thickness of web, $\qquad$ t = 0.25 b = 0.25 × 0.6 d = 0.15 d

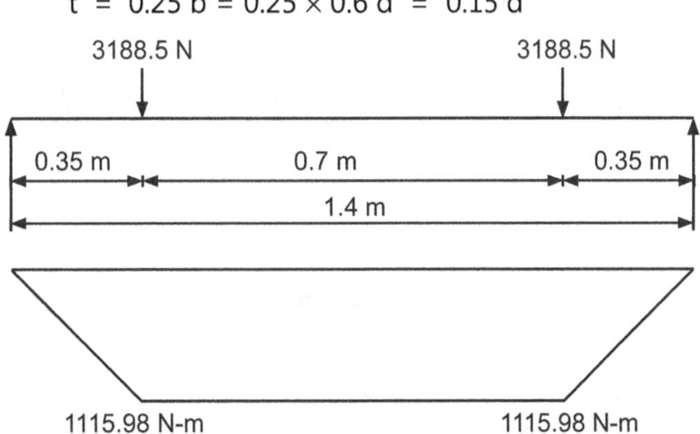

Fig. 5.4 : Simply supported beam with bending moment diagram

Therefore, moment of inertia about N.A. is given by,

$$I = \frac{bd^3 - ch^3}{12}$$

But, $\qquad\qquad c = b - t = 0.6\,d - 0.15\,d \qquad\qquad \therefore \ c = 0.45\,d$

and $\qquad\qquad h = d - 2w = d - 2(0.2\,d) = d - 0.4\,d \qquad \therefore \ h = 0.6\,d$

$$\therefore \qquad I = \frac{0.6\,d \times d^3 - 0.45\,d\,(0.6\,d)^3}{12} = \frac{0.6\,d^4 - 0.0972\,d^4}{12} = \frac{0.5028\,d^4}{12}$$

$$\therefore \qquad I = 0.0419\,d^4$$

Therefore, by using flexural equation,

$$\frac{M}{I} = \frac{\sigma_b}{y}$$

$$\therefore \qquad \frac{1115.98}{0.0419\,d^4} = \frac{88.29 \times 10^6}{\dfrac{d}{2}}$$

$$\therefore \qquad d^3 = 1.508 \times 10^{-4} = \mathbf{0.0532\ m}$$

Other dimensions of I-section are;

Width of flange,	b	$= 0.6\,d$
$\therefore$	b	$= 0.6 \times 0.0532$
$\therefore$	b	$= \mathbf{0.0319\ m}$
Thickness of flange,	w	$= 0.2\,d$
$\therefore$	w	$= 0.2 \times 0.0532$
$\therefore$	w	$= \mathbf{0.0106\ m}$
Thickness of web,	t	$= 0.25\,b = 0.25 \times 0.6\,d$
$\therefore$	t	$= 0.15\,d = 0.15 \times 0.0532$
$\therefore$	t	$= \mathbf{0.0079\ m}$
Also,	c	$= 0.45\,d$
$\therefore$	c	$= 0.45 \times 0.0532$
$\therefore$	c	$= \mathbf{0.0239\ m}$
And,	h	$= 0.6\,d$
$\therefore$	h	$= 0.6 \times 0.0532$
$\therefore$	h	$= \mathbf{0.0319\ m}$

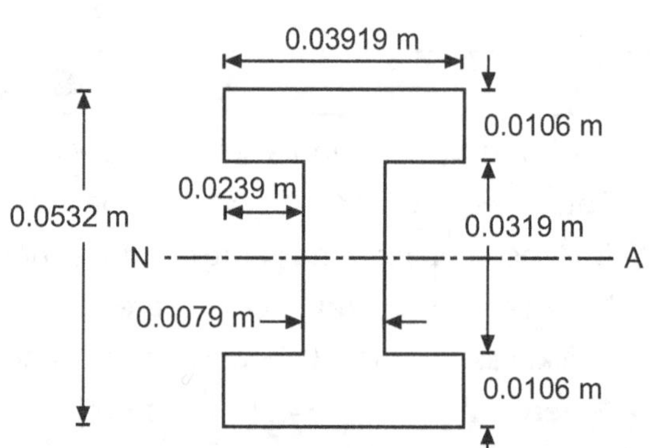

Fig. 5.5 : I-section of axle beam

Problem 5.2 : Design the I-section of front axle of an automobile with the following data :

Total weight of car = 14715 N, Load taken by front axle = 6867 N, Wheel track = 1.5 m

Distance between the center of the spring pads = 0.7 m.

Width of flange and thickness are 0.6 and 0.2 of the overall depth of the section.

Thickness of web = 0.25 of width of flange, Working stress = 88.29×10^6 N/m^2.

Given : $W_{Total} = 14715$ N, Load on front axle = 6867 N, Wheel track = 1.5 m

$\qquad b = 0.6\,d, \quad w = 0.2\,d, \quad t = 0.25\,b, \quad \sigma_b = 88.29 \times 10^6$ N/m^2.

Solution : Load on each wheel of front axle is,

$$W = \frac{\text{Load taken by front axle}}{2} = \frac{6867}{2} = 3433.5\ \text{N}$$

Assuming the axle behave like a simply supported beam as shown in Fig. 5.6.

Therefore, maximum bending moment,

$$M = W \times l = 3433.5 \times 0.35 = 1201.725\ \text{N-m}$$

Assuming overall depth of I-section = 'd' m

$\therefore$ Width of flange,   b = 0.6 d

     Thickness of flange,   w = 0.2 d

     Thickness of web,   t = 0.25 b = 0.25 × 0.6 d

$\therefore$ t = 0.15 d

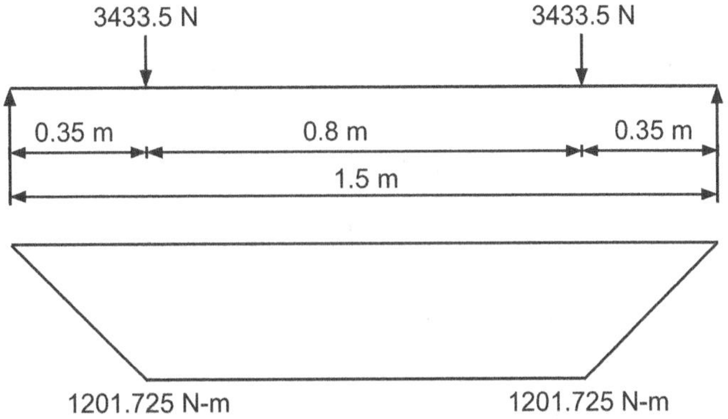

Fig. 5.6 : Simply supported beam with bending moment diagram

Therefore, moment of inertia about N.A. is given by,

$$I = \frac{bd^3 - ch^3}{12}$$

$\therefore$ c = b − t = 0.6 d − 0.15 d $\therefore$ c = 0.45 d

and   h = d − 2w = d − 2(0.2 d) = d − 0.4 d $\therefore$ h = 0.6 d

$\therefore$

$$I = \frac{0.6\ d \times d^3 - 0.45\ d\ (0.6\ d)^3}{12} = \frac{0.6\ d^4 - 0.0972\ d^4}{12} = \frac{0.5028\ d^4}{12}$$

$\therefore$ I = 0.0419 d⁴

$$I = 0.0419\ d^4$$

Therefore, by using flexural equation,

$$\frac{M}{I} = \frac{\sigma_b}{y}$$

$\therefore$

$$\frac{1201.725}{0.0419\ d^4} = \frac{88.29 \times 10^4}{\dfrac{d}{2}}$$

$\therefore$ $d^3 = 1.624 \times 10^{-4}$

$\therefore$ d = **0.0545 m**

Other dimensions of I-section are;

Width of flange,   b = 0.6 d

$\therefore$ b = 0.6 × 0.0545

$\therefore$ b = **0.0327 m**

Thickness of flange,   w = 0.2 d

$\therefore$ w = 0.2 × 0.0545

$\therefore$ w = **0.0109 m**

Thickness of web,   t = 0.25 b = 0.25 × 0.6 d

$\therefore$ t = 0.15 d = 0.15 × 0.0545

$\therefore$ t = **0.0082 m**

Also,   c = 0.45 d

$\therefore$ c = 0.45 × 0.0545

$\therefore$ c = **0.0245 m**

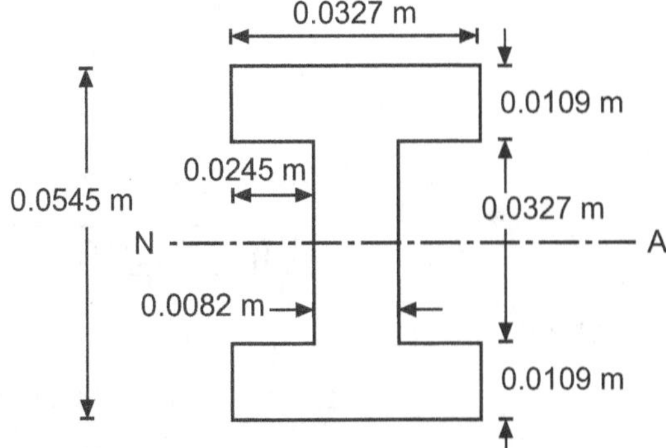

Fig. 5.7 : I-section of axle beam

And, $h = 0.6\,d$

∴ $h = 0.6 \times 0.0545$

∴ $h = \textbf{0.0327 m}$

5.2 DESIGN OF REAR AXLE   (S-16)

Q.1.   *Describe the design procedure of rear axle.*   (W-14)

Q.2.   *Draw a neat and well labelled diagram of fully floating rear axle.*   (S-15)

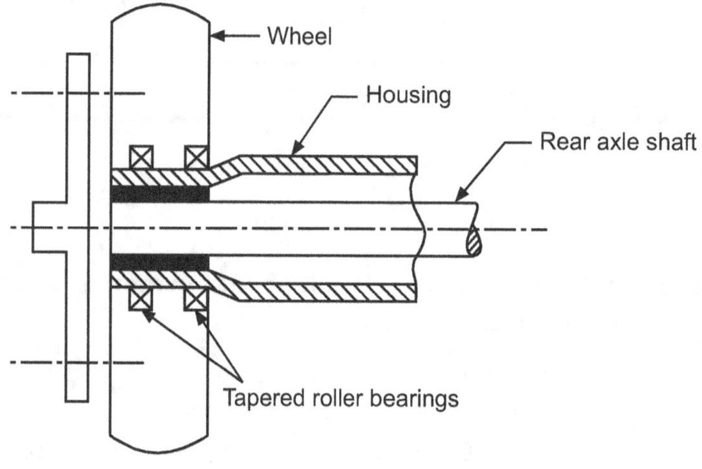

Fig. 5.8

Design Procedure of Rear Axle :   (W-16)

The rear axle is designed on the basis of shaft design, by using torsional equation

$$\frac{T_{RA}}{J_{RA}} = \frac{\tau}{r}$$

where, T_{RA} = Torque transmitted by rear axle shaft

J_{RA} = Polar moment of inertia of rear axle

τ = Torsional shear stress N/mm^2

$r = \dfrac{d}{2}$ (for solid shaft)

$r = \dfrac{d_o}{2}$ (for hollow shaft)

Also, torque transmitted by rear axle shaft,

$$T_{RA} = T_e \times G_1 \times G_d$$

where, T_e = Engine torque

G_1 = Maximum gear ratio in gear box

G_d = Final gear reduction ratio in differential

Power transmitted by shaft,

$$P = \frac{2\pi N T_e}{60}$$

Polar Moment of Inertia (J_{RA}) :

$$J_{RA} = \frac{\pi}{32} \times d^4 \qquad \text{... For solid shaft}$$

$$= \frac{\pi}{32} \times (d_o^4 - d_i^4) \qquad \text{... For hollow shaft}$$

By simplifying above equations, we get

Torque Transmitted by Rear Axle :

$$T_{RA} = \frac{\pi}{16} \times \tau \times d^3 \qquad \text{... For solid shaft (3.12)}$$

Also, $$T_{RA} = \frac{\pi}{16} \times \tau \times d_o^3 (1 - K^4) \qquad \text{... For hollow shaft (3.13)}$$

where, $$K = \frac{d_i}{d_o}$$

$$d_i = \text{Inner diameter of shaft}$$

$$d_o = \text{Outside diameter of shaft}$$

From equation (3.12) and (3.13) above, the diameters of rear axles are calculated.

SOLVED PROBLEMS

Problem 5.3 : *Draw neat sketch of the fully floating axle. And design the diameter of rear axle shaft for fully floating type with following data :*

Engine power = 10 kW at 300 rpm,   Gear box ratio = 4:1, 2.4:1, 1.5:1 and 1:1

Differential reduction = 6:1, τ for the shaft = 70 N/mm^2

Solution : For Fig. : Refer Fig. 5.8.

Given : P = 10 kW = 10×10^3 Watt; N = 300 rpm;

Maximum gear reduction ratio for gear box = 4 : 1;

Differential reduction = 6 : 1;

Shear stress, τ = 70 N/mm^2

1. Torque produced by the engine, T_e

$$P = \frac{2\pi \cdot N \cdot T_e}{60}$$

$\therefore \qquad T_e = \frac{60 \times P}{2\pi N} = \frac{60 \times 10 \times 10^3}{2\pi \times 300} = \frac{600 \times 10^3}{1884} = 318.471 \text{ N-m}$

$\therefore \qquad T_e = \mathbf{318.471 \times 10^3 \text{ N-mm}}$

2. Now, torque transmitted by rear axle shaft, T_{RA}

$$T_{RA} = T_e \times G_1 \times G_d = 318.471 \times 10^3 \times 4 \times 6 = \mathbf{763.304 \times 10^3 \text{ N-mm}}$$

Let us 'd' be diameter of rear axle shaft,

$$T_{RA} = \frac{\pi}{16} \times \tau \times d^3$$

$\therefore \qquad 7643.304 \times 10^3 = \frac{\pi}{16} \times 70 \times d^3 = 13.74 \, d^3$

$$\therefore \qquad \frac{7643.304 \times 10^3}{13.74} = d^3$$

$$\therefore \qquad d^3 = 556.281 \times 10^3$$

$$\therefore \qquad d = 82.24 \text{ mm} \quad \text{Say 83 mm}$$

Diameter of rear axle shaft, **d = 83 mm.**

Problem 5.4 : *Design a rear axle for the following data,*

Engine power = 40 kW at speed 2000 rpm,   Lower gear box ratio = 3 : 1

Differential reduction ratio = 5,   The allowable shear stress = 56 MPa

Solution : Given : $\qquad$ P $=$ 40 kW $= 40 \times 10^3$ Watt

$$N = 2000 \text{ rpm}$$

$$\text{Ger box ratio, } G_1 = 3$$

$$\text{Differential reduction ratio} = 5$$

$$\tau = 56 \text{ MPa} = 56 \text{ N/mm}^2$$

1. **Torque produced by the engine, T_e**

$$P = \frac{2\pi N T_e}{60}$$

$$\therefore \qquad T_e = \frac{60 \times P}{2\pi N} = \frac{60 \times 40 \times 10^3}{2\pi \times 2000} = \frac{2400 \times 10^3}{12560} = 191.08 \text{ N-m}$$

$$= \mathbf{191.08 \times 10^3 \text{ N-mm}}$$

2. **Now, torque transmitted by rear axle shaft, T_{RA}**

$$T_{RA} = T_e \times G_1 \times G_d = 101.08 \times 10^3 \times 3 \times 5 = \mathbf{2866.2 \times 10^3 \text{ N-mm}}$$

3. **Let us 'd' be the diameter of rear axle shaft**

$$T_{RA} = \frac{\pi}{16} \times \tau \times d^3$$

$$\therefore \qquad 2866.2 \times 10^3 = \frac{\pi}{16} \times 56 \times d^3$$

$$\therefore \qquad \frac{2866.2 \times 10^3 \times 16}{\pi \times 56} = d^3$$

$$\therefore \qquad d^3 = \frac{45859.2 \times 10^3}{175.84}$$

$$\therefore \qquad d = 63.89 \quad \text{say 64 mm}$$

$\therefore$ The diameter of rear axle, **d = 64 mm**

Problem 5.5 : *Design fully floating rear axle if :*

(i)   *Engine power = 80 kW at 5000 rpm.*

(ii)   *Gear box ratio 4:1, 2.4:1, 1.5:1, and 1:1, differential reduction 5:1.*

Shear stress for shaft material 65 N/mm². Sketch the arrangement of axle.

Solution : Given : P = 80 kW $= 80 \times 10^3$ Watt, N = 5000 rpm,

G_1 = Maximum gear box reduction ratio = 4, G_d = Differential reduction = 5,

τ = Shear stress = 65 N/mm².

1.　Torque produced by the engine, T_e

$$P = \frac{2\pi N T_e}{60}$$

$\therefore \quad T_e = \frac{60 \times P}{2\pi N} = \frac{60 \times 80 \times 10^3}{2\pi \times 5000} = \frac{4800 \times 10^3}{31400} = 152.87 \text{ N-m}$

$$= \mathbf{152.87 \times 10^3 \text{ N-mm}}$$

2.　Now, torque transmitted by rear axle, T_{RA}

$$T_{RA} = T_e \times G_1 \times G_d = 152.87 \times 10^3 \times 4 \times 5 = \mathbf{3057.4 \times 10^3 \text{ N-mm}}$$

3.　Let us 'd' be the diameter of rear axle shaft

$$T_{RA} = \frac{\pi}{16} \times \tau \times d^3$$

$\therefore \quad 3057.4 \times 10^3 = \frac{\pi}{16} \times 65 \times d^3$

$\therefore \quad 3057.4 \times 10^3 = 12.75 \, d^3$

$\therefore \quad d^3 = \frac{3057.4 \times 10^3}{12.75} = 239796.0784$

$\therefore \quad d = 62.13 \text{ mm}$

$\therefore \quad$ Diameter of rear axle, $d = \mathbf{62.13 \text{ mm}}$　say　$\mathbf{d = 63 \text{ mm}}$

Problem 5.6 : *Design diameter of fully floating rear axle, if engine power is 80 kW at 5000 rpm. Gearbox ratios are 4 : 1, 2.4 : 1, 1. 5 : 1 and 1 : 1. The differential reduction is 5 : 1. Allowable sheer stress for shaft material is 65 N/mm². Sketch the arrangement of the axle.* 　**(W-15)**

Solution : Given :　　　$P = 80 \text{ kW} = 80 \times 10^3 \text{ W}, \quad N = 5000 \text{ rpm}$

Maximum gear ratio,　　$G_1 = 4 : 1$

Differential reduction,　$G_d = 5 : 1$

Shear stress,　　　　　$\tau = 65 \text{ N/mm}^2$

1.　The torque transmitted by the engine, T_e :

$$P = \frac{2\pi N T_e}{60}$$

$\therefore \quad 80 \times 10^3 = \frac{2 \times 3.14 \times 5000 \times T_e}{60}$

$\therefore \quad T_e = \frac{80 \times 10^3 \times 60}{2 \times 3.14 \times 5000} = 152.86 \text{ N-m} = \mathbf{152.86 \times 10^3 \text{ N-mm}}$

2.　Torque transmitted by rear axle shaft, T_{RA} :

$$T_{RA} = T_e \times G_1 \times G_d = 152.86 \times 10^3 \times 4 \times 5 = \mathbf{3057.2 \times 10^3}$$

3.　Let,　　　　　　　$d = $ diameter of rear axle

$$T_{RA} = \frac{\pi}{16} \times \tau \times d^3$$

$$\therefore \qquad 3057.2 \times 10^3 = \frac{\pi}{16} \times 65 \times d^3$$

$$\therefore \qquad d^3 = \frac{3057.2 \times 10^3 \times 16}{3.14 \times 65}$$

$$d^3 = 239662.9$$

$$d = \mathbf{62.11\ mm} \cong \mathbf{64\ mm}$$

Arrangement of fully floating rear axle : Refer Fig. 5.8.

Problem 5.7 : *Design a rear axle for engine power 40 kW at a speed of 2000 r.p.m. Lower gear box ratio 3 : 1 and differential reduction as 5. Take allowable shear stresses 56 MPa.*   **(W-17)**

Solution : Given : P = 40 kW, N = 2000 r.p.m., Lower gear ratio; G_1 = 3 : 1,

Differential reduction; G_d = 5 : 1.

Torque transmitted by the engine T_e :

$$P = \frac{(2 \times \pi \times N \times T_e)}{60}$$

$$\therefore \qquad 40 \times 10^3 = \frac{(2 \times \pi \times 2000 \times T_e)}{60}$$

$$\therefore \qquad T_e = 190.98\ Nm = \mathbf{190.98 \times 10^3\ Nmm}$$

Now torque transmitted by rear axle shaft TRA,

$$T_{RA} = T_e \times G_1 \times G_d = 190.98 \times 10^3 \times 3 \times 5$$

Problem 5.8 : *Design the diameter of rear axle shaft for fully floating type with the following data :*

Engine power = 10 kW at 300 r.p.m., Gear box ratio = 4 : 1, 2.4 : 1, 1.5 : 1, 1 : 1.

Differential reduction = 6 : 1, Shear stress for shaft material = 70 N/mm².

Solution : Given : P = 10 kW = 10×10^3, N = 300 r.p.m., Maximum gear ratio; G_1 = 4 : 1,

Differential reduction; G_d = 6 : 1, Shear stress = 70 N/mm².

Torque transmitted by the engine, T_e :

$$P = \frac{2\pi NT}{60}$$

$$\therefore \qquad 10 \times 10^3 = \frac{2 \times 3.14 \times 300 \times T_e}{60}$$

$$\therefore \qquad T_e = 318.47\ Nm = \mathbf{318.47 \times 10^3\ Nmm}$$

Torque transmitted by rear axle shaft, T_{RA} :

$$T_{RA} = T_e \times G_1 \times G_d = 318.47 \times 10^3 \times 4 \times 6 = 7643.28 \times 10^3$$

Let, $\qquad\qquad$ d = Diameter of rear axle,

$$T_{RA} = \frac{\pi}{16} f_s\, d^3$$

$$7643.28 \times 10^3 = \frac{\pi}{16} \times 70 \times d^3$$

$$d^3 = 556098.64$$

$$d = 82.233\ mm \cong 83\ mm$$

Problem 5.9 : *The rear axle shaft connecting differential to side wheel is required to transmit 40 kW at 1600 r.p.m. If maximum torque is two times average torque and allowable shear stress is 80 N/mm^2 for axle shaft material, find out diameter of axle shaft if (a) shaft is solid, (b) shaft is hollow with outside diameter 1.6 times inside diameter.*

Solution : Given Data : P = 40 kW = 40 × 10^3 W, N = 1600 r.p.m., T_{max} = 2 T_{avg}, τ = 80 N/mm^2

(a) Shaft is solid : Let,   d = Diameter of shaft

We know that, torque transmitted by shaft,

$$T_{avg} = \frac{P \times 60}{2\pi N} = \frac{40000 \times 60}{2\pi \times 1600} = 238.73 \text{ N-m} = 238.73 \times 10^3 \text{ N-mm}$$

Maximum torque transmitted by shaft :

$$T_{max} = 2 \times T_{avg} = 2 \times 238.73 \times 10^3 = 477.46 \times 10^3 \text{ N-mm}$$

We also know that, maximum torque transmitted by shaft,

$$T_{max} = \frac{\pi}{16} \times d^3 \times \tau$$

$$477.46 \times 10^3 = \frac{\pi}{16} \times d^3 \times 80$$

∴   d = **31.20 mm**   say **32 mm**

(b) Shaft is hollow with outside diameter 1.6 times inside diameter :

Let, d_i = Inside diameter of hollow shaft

                         d_o = Outside diameter of hollow shaft = 1.6 d_i

$$k = \frac{d_i}{d_o} = \frac{d_i}{1.6\, d_i} = 0.625$$

We know that for hollow shaft, maximum torque transmitted by shaft,

$$T_{max} = \frac{\pi}{16} (d_o)^3 \times \tau \times (1 - k^4)$$

$$477.46 \times 10^3 = \frac{\pi}{16} (d_o)^3 \times 80 \times (1 - 0.625^4)$$

∴ d_o = **32.97 mm**   say **34 mm**

and d_i = 34 × 0.625 = **21.25 mm**

Problems for Practice

1. Design a suitable I-section for the front axle assuming the following data :
    Total weight of car = 14000 N,   Load taken by front axle = 6550 N,   Wheel track = 1.4 m,
    Distance between the center of the spring pads = 0.7 m.
    Width of flange and thickness are 0.6 and 0.2 of the overall depth of the section.
    Thickness of web = 0.25 of width of flange,   Working stress = 90 × 10^6 N/m^2.

2. Design a suitable I-section for the front axle assuming the following data :
    Total weight of car = 15500 N,   Load taken by front axle = 7500 N,   Wheel track = 1,5 m
    Distance between the center of the spring pads = 0.7 m.
    Width of flange and thickness are 0.6 and 0.2 of the overall depth of the section.
    Thickness of web = 0.25 of width of flange,   Working stress = 92 × 10^6 N/m^2.

MSBTE Questions and Answers (As per G-Scheme)

Summer 2016

1. Describe design procedure for full floating rear axle.   **(4 M)**

Ans. Refer Article 5.3.

Winter 2016

1. Design the design procedure of a rear axle.   **(4 M)**

Ans. Refer Article 5.3.

Winter 2017

1. Design a rear axle for engine power 40 kW at a speed of 2000 r.p.m. Lower gear box ratio – 3 : 1 and differential reduction as 5. Take allowable shear stresses 56 MPa.   **(4 M)**

Ans. Refer Solved Problem 5.6.

Summer 2018

1. Design the diameter of rear axle shaft for fully floating type with the following data :
   Engine power = 10 kW at 300 r.p.m., Gear box ratio = 4 : 1, 2.4 : 1, 1.5 : 1, 1 : 1.
   Differential reduction = 6 : 1, Shear stress for shaft material = 70 N/mm^2.   **(4 M)**

Ans. Refer Solved Problem 5.7.

Winter 2018

1. The rear axle shaft connecting differential to side wheel is required to transmit 40 kW at 1600 r.p.m. If maximum torque is two times average torque and allowable shear stress is 80 N/mm^2 for axle shaft material, find out diameter of axle shaft if (a) shaft is solid, (b) shaft is hollow with outside diameter 1.6 times inside diameter.   **(6 M)**

Ans. Refer Solved Problem 5.8.

✍ ✍ ✍